Perspectives on Risk, Assessment and Management Paradigms

Edited by Ali G. Hessami

Published in London, United Kingdom

IntechOpen

Supporting open minds since 2005

Perspectives on Risk, Assessment and Management Paradigms
http://dx.doi.org/10.5772/intechopen.77127
Edited by Ali G. Hessami

Contributors
Jan Folkmann Wright, Marek Rozycki, Matthew Cook, John P.T. Mo, K Srinivas, Rafiq M. Choudhry, Christopher Gan, Nguyen Thi Thieu Quang, Simon Grima, Frank Bezzina, Riccardo Beltramo, Enrica Vesce, Paola De Bernardi, Paolo Cantore, Sergio Margarita, Carine J. Yi, Tim Park, Jordi Botet, Ali Hessami

First published in London, United Kingdom, 2019 by IntechOpen
IntechOpen is the global imprint of INTECHOPEN LIMITED, registered in England and Wales, registration number: 11086078, The Shard, 25th floor, 32 London Bridge Street
London, SE19SG - United Kingdom
Printed in Croatia

British Library Cataloguing-in-Publication Data
A catalogue record for this book is available from the British Library

Additional hard copies can be obtained from orders@intechopen.com

Perspectives on Risk, Assessment and Management Paradigms
Edited by Ali G. Hessami
p. cm.
Print ISBN 978-1-83880-133-5
Online ISBN 978-1-83880-134-2

For EU product safety concerns:
IN TECH d.o.o., Prolaz Marije Krucifikse Kozulić 3, 51000 Rijeka, Croatia, info@intechopen.com or visit our website at intechopen.com.

Meet the editor

Professor Ali Hessami, FRSA, PhD, BSc (Hons), EurIng, CEng, FIET, SMIEEE, is the Director of R&D and Innovation at Vega Systems. He is an expert in systems assurance and safety, security, sustainability, and knowledge assessment/management methodologies and has a background in design and development of advanced control systems for business and safety critical industrial applications. He chairs the IEEE P7000 Ethics in Technology standard and is the vice-chair of the Ethics Certification Program at IEEE. Professor Hessami is a visiting professor at London City University's Centre for Systems and Control and at the Beijing Jiaotong University School of Electronics and Information Engineering. He is also a fellow of the Royal Society of Arts, a fellow of the UK Institution of Engineering and Technology and a senior member of IEEE.

Contents

Preface

This book explores various paradigms of risk, domain-specific interpretation, and application requirements and practices driven by mission and safety critical to business and service entities. The selected chapters fall into four categories intended to guide the readers with a specific focus on identifying and gaining insight into and value from discipline-specific case studies and state of practice. The contents are classified as follows:

1. Risk, Principles and Foundations

2. Risk, Public/Private and Financial Services Context

3. Risk, Technology and Engineering Context

4. Risk, the Future Paradigms

In an increasingly intertwined global community, understanding, evaluating, and addressing risks and rewards will pave the way for a more transparent and objective approach to benefiting from the promises of advanced technologies while maintaining awareness and control over hazards and risks. This book is conceived to inform decision-makers and practitioners of best practice across many disciplines and sectors while encouraging innovation towards a rational and holistic approach to risk in their areas of professional interest and practice.

For their professional stance, hard efforts, and diligence in reviewing and implementing the proposed enhancements to their manuscripts, I would like to thank all authors who have contributed to this book. We have also incorporated reviewer's recommendations and the book is much improved as a consequence.

Special thanks goes to Dolores Kuzelj (Author Service Manager) for her dedicated support in the submission and reviewing process and her suggestions for further improvement. Finally, all thanks goes to IntechOpen for publishing this book.

Ali G. Hessami
Professor,
Director of R&D and Innovation Vega Systems,
London, United Kingdom

Section 1

Risk Principles and Foundations

Chapter 1

Introductory Chapter: A Systems Framework for Risk Assessment

Ali Hessami

1. Introduction

Throughout the ages, man's preoccupation with determining and controlling his destiny has sparked a keen interest in foretelling the future. This has strangely been based on a linear notion of time and event space, considering the future, a mere extension of the past. Another intriguing facet of this enterprise is probably driven by the maxim that "good news is no news"; therefore, more weight and prominence has been given to negative and downside forecasting, focusing on the detrimental and potentially catastrophic events. This perspective is tacitly echoed in the news media in which a significant proportion of events covered are of depressive, sinister, vile and sometimes tragic pedigree, almost to the detriment of the positivity and hope.

The systematic approach to the understanding and judicious resolution of complex events practiced around the eighteenth century BC by Babylonian Asipus bears close resemblance with the risk-based paradigms in vogue nearly two millennia later. However, we argue that the public and private enterprises need to transcend beyond compliance with regulatory frameworks that typically set baseline benchmarks for acceptable performance and risk of adversity.

The risk management industry in vogue today epitomises this unipolar and adversarial perspective by spending time, resource and effort in predicting and at best avoiding future incidents, accidents and their consequent losses. This is hardly a message of hope, progression and transformation. While addressing foreseeable future adversities is a rational and prudent measure, it lacks the motive force of advancement and success without which we will remain stagnant and at best free from harm or loss.

We present a critique of the obsession with risk and set out a systematic and equitable framework for decision-making, supported by a new methodology for elicitation, representation, communication and resolution of real-world issues and problems. The systematic assessment principles developed here are proposed as a universal set of goals pertinent to assessment of risks arising from all systems irrespective of type, size, origin, environment and function.

2. A question of balance

"Almost every wise saying has an opposite one, no less wise, to balance it."
G. Santayana

The prophet Zoroaster (630–550 BC), born in the mountains east of the Caspian Sea, founded the Persian religion of Zoroastrianism [1]. He is claimed to have received revelations from Ahura Mazda (the "Lord Wisdom") at an early age. The Persian scripture known as the Avesta contains hymns called Gathas, which are attributed to Zoroaster. His teachings portray the universe as a battle ground for good and evil. He also taught about the purpose of living in the world of opposites founded on the premise that there must be an underlying intelligence to the universe and laws governing it. Scholars claim that Zoroastrian doctrine has had a fundamental influence on the subsequent religions of the Middle Eastern origin and through them, on the civilisations founded on these creeds.

Further east, a mystical and intuitive school of thought, Taoism, emerged around the sixth century BC as a reaction to the perceived limitations of rational knowledge. The adherents of this philosophy developed an essentially scientific yet empirical observation of nature, in order to discover the characteristics of reality, Tao, believing that ultimate reality is beyond the capacity of reasoning and rational thought [2]. In this intuitive quest, the Taoist sages came to profound insights about nature, the most important of which are transformation and change. They interpreted the changes in nature as a result of interplay between polar opposites of *yin* and *yang*, which are seen as dynamically linked opposites. This implies an implicit belief in the unity of opposites, which has more affinity with quantum mechanical interpretation of universe than an extension of rational insight. The belief in the continuous interplay of opposites led to two fundamental Taoist rules about human conduct. The first emphasises that to achieve anything, one ought to start with its opposite such as "in order to take, one will surely give first." The second rule states that in order to retain anything, one should admit in it, something of its opposite pole such as "be bent and you will remain straight." In a similar analogy to the Zoroastrian forces of good and evil, the Taoists strive to attain and maintain a dynamic balance between the polar opposites of *yin* and *yang*, which are seen as a spontaneous and innate tendency in all things. In this view, humans should model their behaviour in harmony with nature, driven by intuitive knowledge.

Further to the west, and almost at the same time, Heraclitus of Ephesus in Greece came to the same realisation about the constant transformation of nature [3]. To this, he added a further observation about the cyclic nature of change. Like Chinese Taoist sages, Heraclitus discerned the dynamic interplay of the polar opposites as a unity, a notion now associated with the findings of modern physics.

The three isomorphic visions of reality, emerging from three advanced civilisations around 600 BC, portray a holistic and harmonious perspective on the nature of existence, reality and truth. They epitomise the need for adoption of a more balanced and realistic approach to the understanding, harnessing and management of polar opposites, threats and opportunities, inherent in every facet of life. This is ancient wisdom tantamount to achievement of a dynamic balance in preference over maximisation of gain or minimisation of loss advocated by the pervasive unipolar philosophies of today.

3. The role of creativity

> *"Uncertainty and mystery are energies of life. Don't let them scare you unduly, for they keep boredom at bay and spark creativity."*
>
> *R.I. Fitzhenry*

The nature of creative behaviour and thought has long been debated by psychologists. They broadly agree that such behaviour is distinguished by its novelty and value. The thinking process inherent in creative behaviour is sometimes referred to as divergent since it moves outward from the problem in a variety of directions, potentially leading to many solutions. This is contrasted with convergent thinking, which moves in a straightforward fashion towards a single specific answer [4]. A similar distinction is made by De Bono [5, 6] where divergent and creative thought processes are referred to as lateral and the conventional thinking as vertical.

A range of specific techniques have been developed that may facilitate the creative process. Many of these are founded on the basic principles of creative thinking. However, apart from anecdotal observations, there is little empirical evidence to support their efficacy. Two broad categories of techniques promoting creative thinking are characterised by the nature of the source. Those relating to the cognition of an individual are known as "intra-individual", while creative thought originating from a group of people is referred to as "inter-individual."

The intra-individual techniques promote divergent thinking by breaking or challenging the mental models in an individual and sometimes treating problems as opportunities [7]. The inter-individual or group-based techniques employ the inherent diversity in perspective and mental set to generate a new composite perspective to a problem. Unfortunately, there is no credible theoretical model to support the group characteristics such as composition or size with respect to its creative performance. The optimal size is often quoted as varying between 5 and 7, while the composition should by necessity include members from various stakeholder groups affected by the problem [8]. By far, brainstorming is the best-known technique for group-based creative ideas generation and problem solving. The four key rules of brain storming are as follows:

- No criticism of any ideas allowed
- All ideas including the absurd ones are welcome
- The more ideas the better
- Composite and piggyback ideas are encouraged

While numerous reports have been compiled in support of effectiveness of brainstorming, recent experimental studies have suggested otherwise. The general conclusions being that brainstorming may be useful in some settings but it is not a substitute for individual production of ideas [4]. De Bono [5, 6] also described a technique called "Six Hats" mostly applied to effective management of meetings. In this approach, each hat represents a different perspective on the problem, avoiding conflict and encouraging constructive and conjunctive group thinking.

Whatever the substance and mechanics of creativity, it is considered as a highly effective process for enhanced productivity and achieving differential business advantage. This is particularly relevant to the competitive, rapidly changing and complex problems facing the business environment of today. Creativity challenges the familiar solutions, concepts and strategies for problem solving, which often dominate our thinking, paving the way for novel, high gain and valuable alternatives to come to the fore. The formidable challenges of complexity, inter-relatedness and rapidly evolving issues of today can only be countered through equally potent and penetrating weapon of creativity. Incremental advancement through vertical thinking is no match for the scale and scope of today's tasks.

4. The paradox?

> *"Nature does nothing without purpose or uselessly."*
> *Plato*

Most human endeavours are underpinned by motivation and drivers that are broadly positive and purposeful. These comprise a broad spectrum of activities and tasks ranging from the pursuit of physiological survival needs to higher level attainment, cognitive and transcendental goals. While these pursuits entail expectation of desired positive outcomes, it is inevitable, however, due to the inherent ontologic and epistemic uncertainties that the objectives fail to materialise in part or whole or entirely unexpected often detrimental outcomes emerge instead, the so-called downside risks. So, the gain and loss, hazard and opportunity are intertwined and omnipresent with different likelihoods unless energy and effort is spent on identifying and analysing the relevant scenarios and factors that may potentially impact on the pursued goals. This is the essence of risk-based mind-set factoring in adversity and loss alongside any purposeful positive endeavour, dating back to Babylonians.

In this bipolar reality, mere focus on gain or loss is tantamount to a partial and jaundiced view of the reality that is contrary to prudence and wisdom. This calls for a rational and holistic framework where the potential for loss and gain is identified, evaluated and assessed by the duty holders, balanced and implemented based on insight, awareness and preferences. This is analogous to the Taoist concept of balance between polar opposites, seeking insight on a desirable level of balance.

The best practice standards treat risk as a potential for a gain or a loss, driven by uncertainty. However, the common parlance treats risk as an undesirable outcome that entails harm and loss. To ensure clarity and appropriate treatment for gain and loss scenarios, we propose a systems framework that comprises the following:

- Hazards → risks

- Opportunities → rewards

However, risks are manifestations of what is generally regarded as undesirable or hazardous circumstances, while rewards are the polar opposites. In this context, a hazard is a causative factor to risk and pertains to a condition, object, state or act with a potential to lead to loss, which may entail business/financial, safety or environmental aspects or a combination of these. The opposite concept to a hazard is an opportunity. This likewise is a causative factor for a reward and is a condition, state or act with a potential to lead to some gain/benefit that may entail personal, societal, technological, business or environmental aspects or a combination of these. The likely realisation of a gain arising from an opportunity is regarded as a reward.

This overall framework is shown in **Figure 1** where typically hazards are transformed into a spectrum of risks and opportunities into rewards, respectively. The outcome is the spectrum and scale of risks and rewards that on balance informs the stakeholders in their desired decisions.

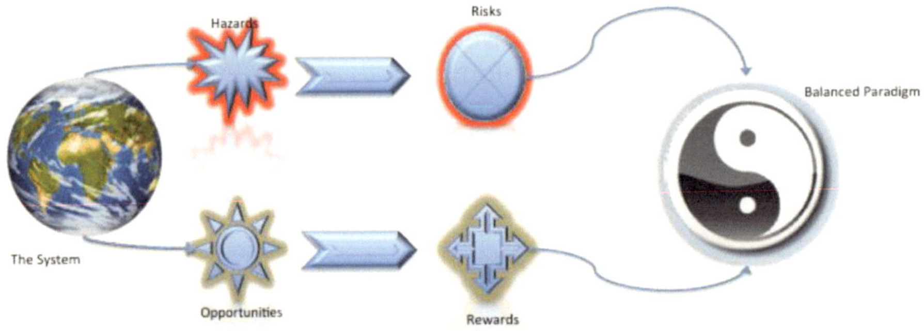

Figure 1.
A holistic risk-reward framework.

Hazards and opportunities are essentially precursors to *risks* and *rewards* and there is a strong argument that they should be identified, assessed and balanced in any rational decision support framework. This framework provides a holistic, clear and unambiguous view of the key influencing factors avoiding confusing *upside* and *downside* terminology often employed to inadequately convey the same concepts or intent.

5. Risk and facets of performance

"To win without risk is triumph without glory."
Pierre Corneille

Before we endeavour to explore the best practice approaches to understanding, assessing and treating risks and rewards, it is constructive to briefly review the facets of a general system's behaviour or emergence that could give rise to areas of concern in terms of potential risks. The following facets of a general system's performance represent generic and often inter-related emergent properties that constitute the focus of attention to realisation of a product, process, service or undertaking:

- Functional and technical
- Commercial
- Environmental/sustainability
- Integrity (reliability, availability, maintainability and safety)
- Security (threat/vulnerability)
- Quality
- Social and personal value (perceived and objective)

Apart from inter-relatedness of the performance facets, some of these emergent properties of products, services, systems and processes are also regulated, which implies risks must be identified and reduced to acceptable or tolerable levels before permit to deployment is given by the relevant authorities [9, 10]. Among these, safety and environment and increasingly security are the regulated facets. The regulations often demand a risk-based process for risk assessment, treatment and demonstration of safe, secure or environmentally friendly performance/behaviour through a documented compliance case [11, 12].

It is also instructive to briefly review the definition and attributes of risk before an attempt is made to develop a structured and systemic framework for its assessment and management.

The ISO key standard on Risk Management Principles and Guidelines [13] describes risk as "Effect of Uncertainty on Objectives" with effect being qualified as a deviation from the expected performance and objective as having aspects such as financial, environmental, health, safety, etc. much akin to the facets of performance elucidated above. The standard also acknowledges that risk is often characterised by events and their consequences and expressed in terms of the combination (product of) likelihood of the occurrence of an event and the consequences of that event.

The IEC Information Security suite of Standards [14–34] adopt the ISO's definition and extends this into the information security domain by adding additional concepts such as threats that can exploit vulnerabilities of an information asset and thereby cause harm to an organisation.

The more recent systems' safety standard for the safety critical railway transportation [35] refers to risk more technically as the "combination of expected frequency of loss and the degree of severity of that loss." However, the standard does not elaborate on the meaning and scope of loss. We define loss as the harm to people (safety including reduction of welfare, injuries and fatalities), environment (damage/contamination/destruction) or detriment to an enterprise (financial/commercial) or a combination of these.

Bearing in mind the definitions, reference to risk is not adequately communicated unless five attributes are specified namely:

Risk = f (C, S, H, L, I) where

- C is the nature of consequence, for example, safety, commercial, etc.
- S is the subject, for example, system, operation, specific people, etc.
- H is the initiating hazard, for example, system crash.
- L is the likelihood/frequency of the consequence arising from the hazard.
- I is the intensity/extent of the loss.

In this spirit, it is not sufficient to express the technical aspect of likelihood of consequence and extent of the loss to transparently and adequately communicate the intent behind a particular risk category.

6. The current best practice

The modern best practice standards [13] advocate a general approach to the assessment and management comprising a set of objectives comprising the following (**Figure 2**):

1. System or context definition
2. Risk identification
3. Risk analysis
4. Risk evaluation
5. Risk treatment
6. Risk monitoring
7. Risk communications

The current deficit is that the best practice standards in risk assessment essentially define a high-level roadmap but do not provide a systematic process supported by essential activities, methodologies and tools to enable the practitioners to

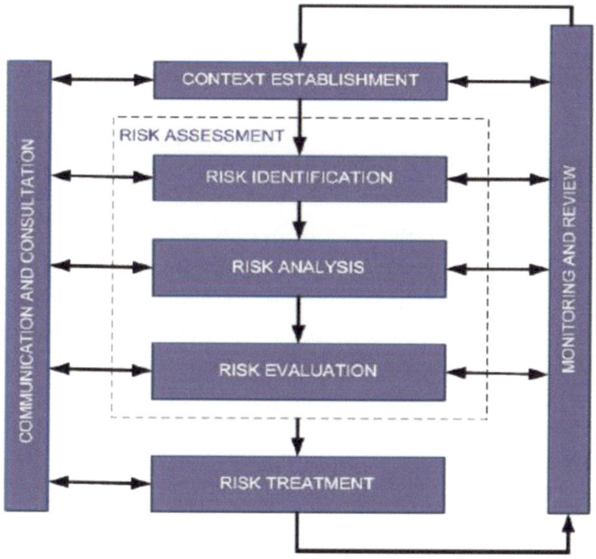

Figure 2.
Best practice risk assessment and management.

implement the requirements in a consistent, comprehensive, verifiable and value-focused manner. Alas, this has led to a plethora of approaches and methodologies that lack credibility, systematicity, systemic rationale and completeness. Most risk assessments tend to be tool centric, that is, in the absence of a principles-based framework and supporting processes, these follow the imperatives of a particular methodology constrained by implementation in the form of a computer-based tool. There is a need for a strategic and systems-based perspective on the requirements for a structured, rational and integrated set of principles that collectively result in understanding, evaluation, assessment and treatment of risks that is not constrained by specific methodologies and associated tools. A candidate solution to this is developed in the following section.

7. The systems framework for risk assessment

> *"First weigh the considerations, then take the risks."*
> *Helmuth von Moltke*

A systems framework for risk assessment constructed on a suite of principles that go beyond the definitions in the best practice standards is developed while also providing guidance on the methodologies and tools necessary to implement each principle. The argument we pose against the risk identification, analysis and evaluation as the sole activities in risk assessment is that risk treatment is fundamentally an integral part of assessment. After sufficient insight is developed in identifying, analysing and evaluating risks against tolerability criteria, many options need to be identified and evaluated to ensure suitable and sufficient risk reduction is achieved. Evaluation of the pertinent options and selection of the most impactful and cost-effective options are integral to risk assessment, treatment and demonstration of compliance with the regulatory requirements.

In the proposed framework, the evaluation, assessment and treatment of risks and rewards can be carried out in a qualitative, quantitative or hybrid manner. Whatever the approach, it is essential, however, that a common currency and compatible outcomes are generated through qualitative or quantitative approaches to facilitate comparison and integration as appropriate.

The systematic approach to the identification, evaluation, assessment and treatment of risks and rewards entails the following principal stages:

- Hazard and opportunity identification within the system context [36]
- Causal analysis (exploring the causation factors for hazards and opportunities) [37–39]
- Consequence analysis (exploring the range of events that potentially arise from escalation of hazards and opportunities)
- Loss/gain analysis (exploring the degree of loss or gain anticipated from predicted consequences)
- Options analysis (exploring viable risk control or reward enhancement solutions)
- Impact analysis (evaluating the expected impact of identified options on risks and rewards)
- Demonstration of diligence and compliance

It is worth noting that the framework of seven principles embraces qualitative and quantitative approaches to the evaluation of risks and rewards to facilitate ranking, judgement and balancing.

The intent, objectives, processes and applicable methodologies for each of the seven principles are beyond the scope of this introductory chapter.

8. The way forward

> *"The universe will reward you for taking risks on its behalf."*
>
> *Shakti Gawain*

The current obsession with risk underwritten by the vast financial and safety sectors portrays an imbalanced perspective on most issues, problems and decisions. While necessary, understanding and management of risks is not singly sufficient to provide a complete basis for rational and realistic decision-making. The emergence of risk-based laws and regulations tends to exacerbate the current myopic view in that risks potentially arising from products, processes, services and systems are subject to legal scrutiny irrespective of the overall contribution to the end users, stakeholders or the society at large.

We have offered a transparent and systematic framework to provide a holistic decision support environment for instances entailing uncertainty and risk. This approach typifies the blend of holism and creativity required to comprehend and tackle the complex and inter-related problems of modern age.

The proposed framework was originally developed by the author in 1997 and was adopted by the UK railway industry's Engineering Safety Management System known as the Yellow Book 2. It became the de facto Code of Practice for risk analysis and assessment in the newly privatised railways in the UK at the time and appropriate training was delivered nationally to all decision-makers in safety and mission critical roles. Later, the framework, process and supporting tools were employed to develop the first national railway quantitative safety risk forecasting model under the heading of risk profiling of railways project in Railtrack plc. Two further variations of the framework were also employed in developing safety risk forecasts for the West Coast Modernisation Programme in the UK and the European Rail Traffic Management System's (ERTMS) safety analysis.

The plethora of risk-based regulations and the underlying principles for tolerability in vogue today should progress towards a more holistic perspective comprising evaluation of hazards and opportunities and assessment of risks alongside the rewards in a given context to provide a more equitable and rational basis for a fair judgement. Mere focus on risk alone provides a myopic view of a more complex systemic reality that goes in the face of rationality, innovation and equity.

Author details

Ali Hessami
Vega Systems, UK

*Address all correspondence to: hessami@vegaglobalsystems.com

References

[1] Bekhradnia S. The Beliefs of Zoroastrianism; 8 January 2007. New Statesman; 2007. https://www.newstatesman.com/uk

[2] Capra F. The Tao of Physics. London: Flamingo; 1986

[3] Kahn C. The Art & Thought of Heraclitus. Cambridge University Press; 1979. pp. 1-23. ISBN: 0-521-28645-X. https://www.cambridge.org/

[4] Ellis HC, Hunt RR. Fundamentals of Cognitive Psychology. Madison: Brown & Benchmark; 1993

[5] De Bono E. Serious Creativity. London: Harper Collins; 1996

[6] De Bono E. Lateral Thinking. London: Penguin; 1970

[7] Rickards T. Creativity and Problem Solving at Work. Aldershot: Gower; 1990

[8] Geschka H. Creative Techniques in Product Planning and Development: A View from West Germany. Source Book for Creative Problem Solving. NY: Creative Education Foundation Press; 1992

[9] United Kingdom, Health and Safety Executive. The Tolerability of Risk from Nuclear Power Stations. Revised Ed. (1992). London: HMSO; 1988

[10] The Official Journal of the European Union. Commission Regulation on a Common Safety Method on Risk Evaluation and Assessment (EC) No. 352/2009; 2009

[11] Hutter BM. The Attractions of Risk-Based Regulation: Accounting for the Emergence of Risk Ideas in Regulation, ESRC Centre for Analysis of Risk and Regulation, Discussion Paper 33; London School of Economics & Political Science; 2005

[12] Condon M. A Tale of Two Trends: Risk-Based and Principles-Based Regulation in Comparative Financial Services Regulation Paper Presented at the Annual Meeting of the Law and Society Association; Hilton Bonaventure, Montreal, Quebec, Canada; May 27, 2008. Available from: http://www.allacademic.com/meta/p236221_index.html

[13] Risk Management Principles and Guidelines. ISO 31000:2009. https://www.iso.org/home.html

[14] Information Technology—Security Techniques—Information Security Management Systems—Overview and Vocabulary. ISO/IEC 27000:2014. https://www.iso.org/home.html

[15] Information Technology—Security Techniques—Information Security Management Systems—Requirements. ISO/IEC 27001:2013. https://www.iso.org/home.html

[16] Information Technology—Information Security Management Systems—Requirements. ISO/IEC 27001. https://www.iso.org/home.html

[17] Information Technology—Code of Practice for Information Security Controls. ISO/IEC 27002. https://www.iso.org/home.html

[18] Information Technology—Information Security Management System Implementation Guidance. ISO/IEC 27003. https://www.iso.org/home.html

[19] Information Technology—Information Security Management—Measurement. ISO/IEC 27004. https://www.iso.org/home.html

[20] Information Technology—Information Security Risk Management. ISO/IEC 27005. https://www.iso.org/home.html

[21] Information Technology—Requirements for Bodies Providing Audit and Certification of Information Security Management Systems. ISO/IEC 27006. https://www.iso.org/home.html

[22] Information Technology—Guidelines for Information Security Management Systems Auditing. ISO/IEC 27007. https://www.iso.org/home.html

[23] Information Technology—Guidelines for Auditors on Information Security Controls. ISO/IEC TR 27008. https://www.iso.org/home.html

[24] Information Technology—Sector-Specific Application of ISO/IEC 27001—Requirements. ISO/IEC 27009. https://www.iso.org/home.html

[25] Information Technology—Information Security Management for Inter-sector and Inter-organizational Communications. ISO/IEC 27010. https://www.iso.org/home.html

[26] Information Technology—Information Security Management Guidelines for Telecommunications Organizations Based on ISO/IEC 27002. ISO/IEC 27011 https://www.iso.org/home.html

[27] Information Technology—Guidance on the Integrated Implementation of ISO/IEC 27001 and ISO/IEC 20000-1. ISO/IEC 27013. https://www.iso.org/home.html

[28] Information Technology—Governance of Information Security. ISO/IEC 27014. https://www.iso.org/home.html

[29] Information Technology—Information Security Management Guidelines for Financial Services. ISO/IEC TR 27015. https://www.iso.org/home.html

[30] Information Technology—Information Security Management—Organizational Economics. ISO/IEC TR 27016. https://www.iso.org/home.html

[31] Information Technology—Code of Practice for Information Security Controls Based on ISO/IEC 27002 for Cloud Services. ISO/IEC 27017. https://www.iso.org/home.html

[32] Information Technology—Code of Practice for Protection of Personally Identifiable Information (PII) in Public Clouds Acting as PII Processors. ISO/IEC 27018. https://www.iso.org/home.html

[33] Information Technology—Information Security Management Guidelines Based on ISO/IEC 27002 for Process Control Systems Specific to the Energy Utility Industry. ISO/IEC 27019. https://www.iso.org/home.html

[34] Health and Safety Environmental Protection—The Offshore Installations (Offshore Safety Directive) (Safety Case etc.) Regulations 2015, Statutory Instrument No. 398, HSE UK. http://www.hse.gov.uk/

[35] BS EN50126-1:2017. Railway Applications, the Specification and Demonstration of Reliability, Availability, Maintainability and Safety (RAMS). CENELEC. https://www.cenelec.eu/standardsdevelopment/ourproducts/europeanstandards.html

[36] Hazard and Operability Studies (HAZOP Studies)—Application Guide. IEC 61882:2016. https://webstore.iec.ch/publication/24321

[37] Fault Tree Analysis (FTA). IEC 61025:2006. https://webstore.iec.ch/publication/4311

[38] Analysis Techniques for System Reliability—Procedure for Failure Mode and Effects Analysis (FMEA). IEC 60812:2006. https://webstore.iec.ch/publication/26359

[39] Ericson II CA. Hazard Analysis Techniques for System Safety, Chapter 22. Wiley; 2005. https://www.wiley.com/

Chapter 2

Decision-making in Risk Management

Jan Folkmann Wright

Abstract

The definition of risk introduced in the ISO 31000 standard of 2009 (2018) is uncertain goal achievement; thus, both negative and positive outcomes can be considered. It also implies that risk is not limited to life and health, but may cover all goals of a company. Risk management thus becomes a question of achieving and optimizing multiple goals. Since safety is but one of several considerations, safety may lose out to other more easily measured objectives of a company, such as economics and compliance with regulatory requirements. Risk analyses have a long history of quantification, a tradition that for various reasons has waned and should be revived if safety goals are to be treated together with other goals of a company. The extended scope affects not only company owners and employees but also neighbors, the local community, and the society at large. The stochastic nature of risk and the considerable time lap between decisions and the multiattributed consequences implies that managing risk is exposed to cognitive biases of many sorts. Risk management should be based on a quantitative approach to risk analysis as a protection against the many cognitive biases likely to be present, and managers should be trained to recognize the most common cognitive biases and decision pitfalls.

Keywords: risk management, decision-making, cognitive bias, behavior analysis, hazard adaptation

1. Introduction

Accidents happen, in the past and present, and efforts to analyze how to avoid their reoccurrence have always been the backbone for improvements in safety. Through the study of the causes and statistics of accidents, their frequency and consequence severity have been reduced. Analytical algorithms and tools were developed, mainly after WW2, supplementing the safety improvements of accident investigations. The analytical approach has evolved considerably over the years in terms of improvements in methodology and calculation capabilities. The evolution has also been a response to the extensions in the scope of both risk causes and consequences, i.e., goals.

Some of the mathematics and statistics of risk were developed to meet the need to decide the average remaining lifetime to estimate the cost of life insurance policies [1]. Practical risk reduction knowledge has accumulated since then in high-risk industries like shipping, chemical plants, oil and gas, nuclear power plants, aviation, and space exploration. Risk was defined in relation to unwanted consequences, as a function of the probability with which an event may happen and how severe it might be (see [2] for an overview). If the causes of risk are known and probability data exist, risk can be calculated in quantitative risk analyses (QRAs). Making decisions based on the results of risk analyses in a systematic way inspired the concept of risk management, with the aim to reduce risk based on findings from QRA. The quantitative nature of this approach made cost-benefit analyses possible. If properly carried out, the result was a better utilization of limited resources, be it money, experts, or other means.

The different applications of risk management in insurance, finance, and industry were developed with little mutual exchange between them. The risk management tradition of finance looked at risk including both gains and losses because of investments, while in industry and engineering, risk was associated with potential loss only. Because risk is an expression of events that may happen in the future, risk is intrinsically uncertain. The decisions that may trigger such events are often made to achieve multiple goals, e.g., profit while maintaining safety margins related to health and environment. The question of how to balance several goals is not trivial. Some might be in conflict, others might support each other. There can be different stakeholders affected by the decision, with different priorities and power of influence, and they might be involved directly or indirectly. The stakeholders can be owners, employees, neighbors, politicians, NGOs, or competitors. Some goals might be certain and others uncertain. Some of the effects of decisions can happen in some distant future, raising the issue of discounting. Since humans' discount is differently than "econs," the rational utility maximizing economic man, the stage was set for differences in opinions and priorities. Decision-making in risk management is therefore a practical application of judgment under uncertainty, a research field developed by Tversky and Kahneman [3, 4] leading to the study of cognitive biases and becoming the foundation for behavioral economics [5].

The definition of risk has undergone major changes, from the product of the severity and probability of unwanted events to uncertain achievement of multiple goals, as reflected in the ISO 31000 "Risk Management," a guideline developed for risk management systems [6]. When the scope is lifted to include the whole company and all its objectives, the concept of enterprise risk management (ERM) is used. In parallel with the "engineering" approach, the auditing and accounting professions have developed an approach to ERM under the COSO label [7] with emphasis on fraud prevention and audit of accounting. Comprehensive systems on how to reduce risk to an acceptable level on a continuous basis are commonly described as Safety Management Systems (SMS), reflecting a broad approach including risk analyses, safety assurance, incident investigations, safety inspections, and audits. In aviation, SMS includes the evaluation of incidents with respect to quality the remaining barriers as well as safety issues that may require a more detailed risk analysis [8].

Concurrent with the development of SMS, vetting systems have emerged as background checks of both people and systems. Vetting is a case-based inspection used by a diversity of institutions, from public agencies in border control to oil majors in relation to suppliers. When an oil tanker is nominated to a charterer and considered for lifting cargo at a terminal which requires the consent of an oil major, the oil major will "vet" the vessel, i.e., inspect and approve the vessel for visits to that terminal. This is usually regarded as a more critical inspection than the internal audits performed by the shipowner because the consequence of a failed vetting is a loss of business. SMS and vetting systems complement each other as the former is a continuous and systems-based approach, while the latter is more detailed and adapted to a practical case.

The different definitions of risk and approaches to mitigate risk may have both a positive effect and a negative effect. On the positive side, competition can lead to improvements in achieving results at a lesser cost. Negative effects can be unnecessary activities and conflicts between the various safety assurance actors, with more bureaucracy and higher costs than necessary. A short history of how risk management emerged will be presented before possible improvements in risk management are discussed. The focus will be more on the practical and less on the theoretical aspects and more on management challenges and less on risk calculation issues.

Deliberate actions to harm, like sabotage and terror, are not covered, although some of the considerations made might be of relevance to acts of terrorism as well.

2. A brief history of risk management

2.1 Approaches to risk reduction

Risk causes included in risk analyses have increased in scope; from an initial focus on technical failures and extreme environmental conditions via operator errors to include problems originating in the culture of organizations and (lack of) management commitment. The trend to include a wider causal picture came gradually during the second half of the last century, motivated by the results of investigations of some spectacular major accidents, like the Bhopal gas disaster in India in 1984 [9], the space shuttle accident of Challenger in 1986 [10], and the capsizing of the roro ferry "Herald of Free Enterprise" in 1987 [11]. Major accidents also occurred in finance, like the bankruptcies of Enron in 2001 and Lehman Brothers [12] in 2008, just to mention a few well-known cases. Concurrent with the extension of the scope of causes, the range of risk consequence has widened and includes effects on the environment, social responsibility issues, and company reputation.

The extensions in scopes of both causes and consequences have made risk assessment more challenging as methods for quantitative risk analyses have not caught up with the increase in scopes. Software tools have made risk management easier as far as the "bookkeeping" of risk status and mitigations is concerned. The quality of the content of the risk registers is however another question, because the extension in scope has come at a cost. The "softer" causes and consequences are usually not quantified since cause-effect relations are difficult to identify and estimate. Authorities require risk assessments of new endeavors and major changes, to be followed up by the industry, although there is shortage on both proper methodologies and qualified analysts. One compromise to this issue was to relax the requirement for quantitative results, if not in theory so in practice. QRAs were no longer behind the wheel and were moved to the backseat of risk management. The lack of numeric rigor in expressions of risk opened a Pandora's box of more subjective assessments. The result was a considerable growth in the number of accident investigations and risk assessments whose quality is dubious. It is difficult to judge whether this development was for the good or the bad, as even a poorly conducted qualitative risk analysis could produce interesting findings; at least risk workshops made people talk to each other over the border of department silos, thus improving in-house communications on risk issues.

Improvements in calculation methods, more powerful computers, and software may help to bring QRA back on track by making it possible to apply advanced modeling techniques capturing both stochastic aspects as well as the dynamic properties of evolving systems. The systems are "hybrid" in the sense that they consist of both technical and human parts, the combination of which raises a principal challenge as to how events taking place in the two are to be explained. This is not a new challenge, as it also was the case with man-machine systems and control room operators. Physical phenomena are explained through cause-effect relations, as are also human actions, and work well for simple cases. The ability to predict future states of physical systems is however reduced when complexity increases, e.g., in forecasts of weather. Prediction becomes even harder when the systems contain humans supposed to make multiattributed decisions, as in politics and economics [13], and one might add, as in risk management.

One remedy is to improve the utilization of knowledge from behavioral and cognitive science in the decision-making part of risk management. Behavior science is relatively young compared to physics and engineering. It is regarded as "softer" in nature and harder to quantify. Of more concern is that the different "schools" in the social and human tradition might give an impression of fragmentation, as disagreements rage between different professions and disciplines. This situation is real, unfortunately, and the main differences are related to what the core subject of behavior science should be and how to explain the phenomena studied. The approach promoted in this chapter is psychology as a natural science of behavior. The explanatory concepts are like those used in evolutionary biology, variation, selection, and replicators, to explain fitness in adaptation rather than cause-effect relations between the mind and body. Like natural or Darwinian selection explains phylogenetic evolution (genetics), so is behavior explained by the selection effects of the consequences of behavior during the lifetime of an individual (ontogenetic evolution) [14].

In the following, a historic review of how the origin of risk has been investigated and understood will be described before possible improvements to the decision-making part of risk management is discussed.

2.2 Accident investigations

An old saying states that fools never learn and the smart ones learn from their own mistakes, while the wise learn from others' mistakes. In other words, improvement starts with efforts to prevent the unwanted event from occurring again, by observation and learning from own or others experience. The key is to identify and understand the causes that made the accident happen to prevent it from happening again. Although this was a reactive approach, over the years the gain was huge. Expressions like the anatomy of accidents and unsafe acts were introduced. Earlier, once a human error was identified, the analysis was believed to be complete, a misconception that could easily lead to a search for scapegoats. Without an understanding of what caused behavior failures, the search for human errors implied to find the responsible individual. Often, this was the man at the end of the chain of causes, the last contributor before the accident. As situational factors were understood as the main causes of human behavior, unsafe acts were considered the result of lack of training, time pressure, man-machine interface design, and other error-prone situations, and human errors could be reduced if precautions were taken [15].

Investigation boards covering several industries were established as national agencies in many countries. Accident causes were categorized as independent or contingent on other events, and as the physics of accidents were better understood, the logical relations between the events, their timing, and sequence leading up to the accident could be described more accurately. Unsafe acts no longer were the sole precursor to accidents, and a more complete causal picture appeared. When a human error was identified that had contributed to an accident, that signaled the start of the analysis, not the end. The chain of causes was further expanded when the investigation moved away from the immediate proximity to the operator and to the functions of the organization, management, owners, and the way the regulatory authorities operated. This extension of the causal scope was undoubtful of value, as people higher up in the organizational hierarchy could influence risk much more than the person at the sharp end of the line.

The change in focus from those executing operations to the designers and planners in management reflects the advance in technology which was about to change the primary human role in work life from manual labor to planning and decision-making. Management, organization, and culture were firmly included as topics to be addressed in accident investigations in most countries by the turn of the century. Aviation can serve as an example of successful accident investigation history. Flight anomalies and pilot error reporting are mandatory for both airliners and ground control. The pilots and air traffic controllers filing incident reports are not subject to legal prosecution unless there is a case of deliberate and serious misconduct. The fact that air flights are possibly the safest transportation means of all when exposure is measured per unit of distance and not per time unit is largely due to lessons learned from nonpunitive reporting systems and findings from thorough accident investigation efforts.

2.3 Technical risk

As accident investigation was established as the primary way of enhancing safety, accident causes were initially understood to be technical and human failures. Improvement both in the reliability of components and in how they were combined in systems resulted in fewer accidents. As reliability theory and calculation methods and tools became available to the engineering community after WW2, QRA methods were developed [16]. Techniques of a more qualitative nature were also developed, like failure mode and effects analysis (FMEA), hazard and operability analysis (HAZOP), and various barrier analyses. These new proactive analytical tools made it possible to improve safety before accidents happened and proved to be an important complement to reactive techniques like investigations of accidents and incidents.

The Reactor Safety Study [17] was probably the first "total" quantitative risk analysis (QRA), also called the Rasmussen report or WASH-1400. The report was published in 1974 after 3 years of work involving more than 50 contractors at a cost of about 4 million USD, equivalent to about 30 million today. The analysis was based on a system reliability approach, where component failure rates were combined using Boolean logic, represented graphically as logical gates in fault trees. The objective of the WASH-1400 study was to calculate a realistic estimate of the risk posed by nuclear power plants as a response to public claims that this new way of producing energy was very dangerous. The study concluded that it was about 1 million times more likely that car driving would be fatal. The study was criticized, partly because the nuclear risk was calculated, while the comparative risks, e.g., from traffic accidents, was based on statistics of real events [18]. The most influential result of WASH-1400 study was that it served as a recipe for similar analyses in other industries, e.g., the offshore oil and gas exploration in the North Sea.

One main reason for the early popularity of risk analyses was that the fault and event tree modeling approach was scalable to any plant type and size, if design drawings, P&ID, and component failure rates were available. The QRA made it possible to include the human as a system component that could fail, like a valve, a pump, or a vessel. In this way, the stage was set for the development of human reliability assessment methods that could feed human error probability data into the system reliability models.

The practice of applying risk analysis methods spread to other sectors and industries. Environmental impact studies were prepared built on the same logic.

Consequence assessments were required before approval of large-scale industry and real estate development projects. Some years passed, however, before risk analyses became a required part of safety work in aviation. One probable reason for this late start is that accident and incident investigations had become quite advanced and were used to a large extent in aviation, providing ample evidence for their positive contributions to flight safety. As the saying goes, don't change a winning team. The various safety methods are however better considered as elements of a broader safety effort, each contributing in their own way to improvements. Risk-based SMS are now mandatory for airliners, airport providers, and air traffic control service providers.

2.4 Human risk

Because the systems that failed also needed humans for operation, maintenance, and repair, human reliability became part of QRA. Assessment methods for human reliability for industrial and defense applications with high potential for major accidents were developed. One early example is control room operations in nuclear power plants [19]. Human tasks and their error probabilities were modeled using event trees like THERP, and tables of human error probabilities were published in a handbook for use in risk analyses [20]. Human errors could be omissions or commissions, meaning that something was forgotten or a wrong act was carried out. Later versions of human reliability models stated human error probabilities as a function of performance shaping factors (PSFs). The models were calibrated using data from experiments, statistics, and expert judgments. Examples of PSFs are quality of the man-machine interface, violation of stereotypes, too high or too low stress level, isolated acts, conflict of motives, quality of feedback, etc. [21, 22]. The human error models were mainly motoric tasks or simple decisions related to the execution of the tasks.

Safety research programs were nurtured by the growth in the British and Norwegian oil and gas offshore activities. In Norway, an increase in safety funding became available after the Aleksander Kielland accident in 1980 where a capsized floatel resulted in 123 fatalities. The Piper Alpha accident on the British sector in 1988 also served as a boost for increased safety efforts, resulting in the safety case approach [23]. The state safety funding in Norway was mainly devoted to occupational safety, workplace democracy, and socio-technical issues, while means to develop human reliability lessened. This was possibly due to pressure from labor unions who exerted considerable influence on the governmental financing of safety research. The focus on the worker as a contributing factor to risk was not politically acceptable, even though situational factors or PSFs were modeled as human error causes. When empirically based failure rate repositories were developed [24], and human error was included in the equipment failure rates, the need for human reliability data vanished, and the human reliability profession was history. In Britain the situation was better due to the larger industrial sector and cooperation between industry, universities, and consultancies [25].

The way humans contribute to risk ranges from simple motoric tasks to complicated decisions that include other people and other institutions. The former was developed quantitatively as human reliability, while the latter, decision failures, have so far not been formally included in QRA to this author's knowledge. Decision errors have however been extensively studied by behavior science in the cognitive bias tradition. The absence of an analysis of decision failures in risk management is probably related to lack of empirical data, the high complexity of decisions under risk, and the shortcomings of behavior science in this area. Decision behavior is by no means understood sufficiently, although progress during the last couple of decades has been significant, as exemplified by cognitive bias research and the policy relevant "nudge" tradition [26]. Machine learning, Bayesian network, and self-learning AI robots are promising research disciplines. For now, a closer look at a few cognitive biases relevant for risk management will have to suffice.

2.5 Financial risk

Economics is probably the field of human endeavor that has been most concerned with risk, covering uncertain outcomes of both positive and negative values. New challenges appear as robots and artificial intelligence (AI) are being applied in finance and trading of stocks and derivatives. It is interesting to observe that AI algorithms use operant selection in AI self-learning, adaptive systems. The rapid innovation is a challenge for regulations because regulatory requirements usually are lagging new technology.

Angner defines economics as "the manner in which people make choices under conditions of scarcity and the results of those choices for society at large" [27]. In Anger's textbook Adam Smith is considered the founder of modern economics and author of influential books like *The Wealth of Nations* and *The Theory of Moral Sentiments*. Smith regarded the economy as a self-regulating system where the price mechanism would balance the supply and demand and thus result in the best allocation of scarce resources, aided by competition in the market and humans driven by self-interest. The self-governing system would reduce the need for a supreme regulatory power, being it the state or the church. The idea that liberty and individual freedom with a minimum of regulation would lead to prosperous outcomes for all members of society can be traced back to the Age of Enlightenment, a movement in Europe during the eighteenth century essential for the opposition against religious and feudal governing of people and commerce. The core of the economic system was the rational, utility-maximizing economic agent or "econ," whose behavior was considered both as normative and descriptive. These assumptions were to be criticized from political and behavioral points of view, respectively.

The economic liberty and individual freedom resulted in a much higher productivity and thus accumulation of wealth, but not for all. The politically based critique of the self-regulating economy leading to prosperity was based on the resulting skewed distribution of the new wealth. This controversy still exists and fuels the conflict between capitalist and socialist ideas on governance of a society and attitudes to market economy.

The critique from the behavioral side was based on research showing that most people did in fact not behave like "econs." An important contribution to understanding how decisions were made beyond the rational-agent concept was the work of Herbert Simon [28] on bounded rationality. When the article "Heuristics and Biases" by Tversky and Kahneman [3] was published in *Science*, human decision failures defined as deviations from the choices of an econ became a prime subject of psychological experiments. Prospect theory was published in *Econometrica* [29] by the same two authors a few years later, formalizing the basis for cognitive decision research. Although the lack of a clear definition of a "bias" was pointed out [30], the research on cognitive bias and heuristics flourished. The impact reached far beyond the field of psychology, as the empirical foundations of neoclassical economics came under attack (see Thaler [5] for a historic overview of the emerging field of behavioral economics). The Nobel Memorial Prize in Economic Sciences was given to Simon in 1978, to Kahneman in 2002, and to Thaler in 2017 as a recognition of their contributions.

That the assumption of the rational agent, economic man or "econ" for short, was disputed did not imply that the rationality of economics as such was rejected.

Economic man was assumed to be the normative case of how decisions should be made if the goal was to maximize outcome for the decision-maker. The prescription fits the stereotype of an omniscient business executive doing transactions in a commercially competitive society. It is unclear why leading economists proposed the rational utility-maximizing agent as a generic, descriptive model for human decision behavior, thus confusing the descriptive with the normative. One reason could be to strengthen the legitimacy of economics as a science based on rational humans, to be backed further by the most rational scientific discipline of all, mathematics. Another reason for the misperception of normative and descriptive could be the lack of interest shown by psychologists, including behavior analysts, in decision-making in business and industry. The "invasion of economists" into the land of the social and behavior sciences might explain some of the skepticism toward economists from behavior scientists. The resentment between the two disciplines might have been strengthened because economists are preferred as managers and administrators, especially in the commercial private sector, and few with a background in behavior science seek such employments or are preferred as candidates.

The Enron case has been portrayed as an example of a major financial risk and an example of willful corporate fraud and corruption, and it led to the dissolution of the Arthur Andersen accounting firm. The various financial disasters that followed were probably not committed conscientiously, as the human's capability to self-justification seems limitless. The Enron case was a major motivation for the Sarbanes-Oxley Act of 2002, leading to much stricter accounting rules. Paradoxically enough, it also led to a burst in the business of accounting companies who then later was delivering the services needed for companies to comply with the new rules. A better alternative would have been to introduce quantitative risk management methodologies developed in the engineering domain, and adapted that to finance, rather than to enforce stricter auditing philosophies that so far had been proven insufficient. The mitigation of finance risk has therefore proceeded more as a kind of compliance management rather than risk management.

2.6 Organizational risk

The change in production technology from manual labor, via mechanization to automation, resulted in more management type of activities like designing, planning, and decision-making. As a reflection of this development, a sociological perspective on risk was introduced to explain why accidents occurred [31]. Concepts like "normal accidents," "an accident waiting to happen," "tightly coupled systems," and "interactive complexity" were introduced to describe the vulnerabilities of high-risk companies. This new understanding was first applied in accident investigations, and the stage was set for a search for causes to risk in the way the organization, i.e., management, was prepared for, or rather was not prepared for, safety. There is however little agreement on how an organization should look like for operations to be safe. Studies of the so-called high reliability organizations (HRO) might give an indication [32], but their way of conducting operations would hardly be accepted as role models for the industry in general.

The Swiss cheese metaphor was introduced to illustrate defense in depth [33] and is a visual representation of how barriers can fail simultaneously, visualized as slices of cheese with holes that are lined up. This was a failure situation that for decades had been modeled in reliability engineering by fault trees with the more precise Boolean and gates. The sociological perspective got a strong foothold in accident investigations but did never make it to the QRA teams other than to visualize and illustrate. A new metaphor is not necessarily an improvement, especially not when attention is diverted from logics and calculation. The sociological perspective on accidents reinforced a qualitative approach to the study of organizational risk causes.

As far as the human contribution to risk is concerned, the change resulted in a move in responsibility from the operator to management and owners. The sociological view on causes to accidents, helped by the auditing focus on compliance with regulations, could non-intentionally lead to the blame and shame culture being lifted from the shop floor to the board room.

The sociological perspective on risk came with a political flavor as management could be considered as potential culprits causing accidents. It put the focus on commercial pressure, reduced manning, budget cuts, insufficient training, and if management failed as role models for safety, by paying lip service to safety priorities while acting otherwise. Not walking the talk was a sure trail leading to a depraved safety culture. This critical attitude toward private enterprise and business is understandable given the safety scandals of the time, as the case of the Pinto can exemplify.

3. Risk management improvements

The story of the Ford Pinto illustrates two interesting issues of relevance to risk management and decision-making. The first issue is the low priority given to safety at the time, and the second one is how animosity toward setting a monetary value on life can hamper safety improvements. The priority issue has been rectified as safety is now given a much higher importance, partly because of negative press coverage and lawsuit compensations. The last issue is related to a reluctance to set a monetary value on life and is still controversial. These two issues need some explanations to bring home. It should be noted that the presentation given in the following is based on the report prepared by the Mother Jones magazine [34].

3.1 The issue of safety priority

During the 1960s the American car industry was met with fierce competition from European and Japanese manufacturers who were targeting the lucrative small-car market. The response from the Ford Motor company was the Pinto, a subcompact car that was put into production at record time. This was achieved by concurrency in engineering design and production of assembly line tools. The consequence was that design changes would be costly if they required any change in manufacturing tools. As the first cars were manufactured, collision tests revealed a serious safety issue related to the position of the fuel tank. It would easily burst by the impact from a rear-end collision, even at low speed. The car would be engulfed in flames if the gasoline was ignited, which was likely to happen because the impact itself would produce sparks. Another safety flaw was that the doors would be jammed at a moderate collision speed, rendering escape impossible. Ford management knew about the design flaws, but nothing was done about it although cheap measures were possible. Safety was not given priority, and money was rather spent on lobbying against safety regulations that were being prepared for the auto manufacturers. More than half a million cars were produced each year, making a huge profit for the Ford company. The number of rear-end collision fatalities has been estimated to be in the range of 500–900 during the 8 years before Ford finally incorporated safety improvements.

The many fire accidents caught the interest of the Mother Jones magazine. Several hundred reports and documents regarding rear-end collisions were studied, including the tests made by the company itself. It was also revealed that an internal memo sent to senior management had compared the cost of redesign of the hazardous position of the fuel tank with the off-court settlement cost of humans that would suffer from accidents. Applying the value of life provided by the National Highway Traffic Safety Administration (NHTSA) [35], the memo concluded that Ford would save almost $70 million by allowing accidents to occur. Mother Jones published the story in their August 1977 issue, and the reaction was devastating. Criminal charges and lawsuits were made, all Pintos were recalled, and the Ford Motor company got some of the worst press an American car manufacturer has ever received.

Although traffic safety had improved over many years [36], it was the investigative journalism by the Mother Jones magazine, and the following attention of the press in general, that made a whole industry set higher priority on safety.

3.2 The issue of the value of life

A cost-benefit analysis applied in the management of safety risk will require a monetary value of human life. The benefit from improved safety is calculated as fatalities and injuries avoided. The costs are mainly due to the mitigative measures and production loss needed for the implementation of the measures. A QRA estimating the risk reduction effect of the mitigation may provide the benefit value, if the values of life and health are given. This way of thinking is considered cynical and calculating by some people. Not doing cost-benefit calculations is for quantitative risk analysts equal to missing the opportunity to save lives.

Safety competes with all other objectives of a company, and the easiest to calculate are economics, production logistics, and marketing. Management must be convinced to initiate safety measures for unacceptable risks. The most cost-effective mitigation measures can be identified if a QRA is prepared. The CEO of a company has usually no training in safety and QRA methods, and neither has any of the other directors nor vice presidents.

The Pinto story is not unique, and the car manufacturing industry might not be the worst. A possible side effect of the media focus and lawsuits to track down the responsible individuals or company can have strengthened a reluctance to apply risk analyses to improve safety, because it is always possible that no-cure-no-pay lawyers could use the results in future lawsuits. A verdict in disfavor of a company is more likely if there is a reason to believe that the risks were known by management, even if being at acceptable levels.

It is likely that the Pinto and other cases that made the headlines prepared the ground for the introduction of the concept of the "amoral calculator company," which is a way to describe different types of business firms on how they would respond to safety regulation and enforcement [37]. The amoral calculator type of companies and management was assumed to be mainly driven by self-interest and profit maximization, assuming they calculated costs and benefits in relation to safety measures to see what they could get away with. An ill-fated consequence of this kind of thinking is that doing cost-benefit calculation of safety measures by itself could be considered an amoral act. This is detrimental to safety because it will make the identification of the most cost-effective safety measures difficult.

4. Behavior analysis and risk management

Simply stated, risk management implies making decisions to influence risk in a predicted and controlled way. The expression above rests on the following criteria: knowledge about the risk level and safety margins are available, and the decision-makers are trained to observe and obey the risk acceptance limits. There are however several reasons why this may not happen: first, QRAs may not be carried out, so there is no reference, i.e., no quantitative measure of the current risk or trend. Second, risk acceptance limits are not defined, so if the risk was known, there is no knowledge of it being too high. Third, decision-makers are not trained to observe and act based on trends in the risk level, relative to risk acceptance limits. And finally, human decision behavior is vulnerable to a range of cognitive biases involving thinking and emotions. Failures to meet the assumptions above can result in a faulty risk management process. The most relevant cognitive biases for decision-making in risk management will be described in the following.

4.1 The base rate neglect and exposure

One of the best known cognitive biases is base rate neglect [3] which occurs when background information is disregarded and the decision is based on superficial and less relevant information. Risk is expressed relative to exposure, like the number of events or incidents divided by the number of opportunities for incidents to happen. When catchy and stereotypical descriptions dominate or replace base rates, decisions may be based on deceptive heuristics reflecting these stereotypes. The base rate neglect bias is especially relevant to qualitative risk estimations because this type of risk analysis does not require quantitative exposure data. The result can be hazard adaptation, an unnoticed slide toward a more lenient risk acceptance behavior. Prior to the faulty decision to launch the space shuttle Challenger in January 1986, the decision process failed on several of the criteria mentioned above. The outcome was the loss of the shuttle and the lives of seven crew members. Better knowledge of QRA and cognitive biases in decision-making under risk and uncertainty might have changed the fatal decision and avoided the accident [38].

4.2 The optimism bias and variation

The optimism bias can be described as a general overestimation of our performance in our favor. In a review of biases [39], it was concluded that optimistic illusions are the only group of misbeliefs that might be adaptive. The optimism bias is associated with harder and longer work periods, which may account for higher pay and promotions [40]. The optimism bias is also associated with an optimistic view of future events and an increased will, and thus ability, to predict and control future outcomes. In the inverse situation, as when prediction and control is not possible, the result is reduced ability to learn. The term "learned helplessness" was coined by Seligman to account for these effects [41].

The optimism bias or overconfidence is probably one of the most common and strongest human fallacies [42]. The bias is found in many different countries and cultures. Examples of the optimism bias are that we engage in more new ventures, establish new relationships, buy lottery tickets, etc., in areas where the expected benefit is much lower than the effort invested [43]. Many new businesses would probably not have been started and inventions not made, unless the effort required was underestimated. Many more activities are started than a realistic and rational utility-maximizing agent would initiate, making the optimism bias the mother of variation and innovation.

Translated into evolutionary terms, the optimism bias fuels the variation upon which selection operates. This is probably the case for both types of evolution: phylogenetic as in genetic inheritance and ontogenetic as when operant behavior is selected by the consequence it produces.

Overconfidence is rampant; we are all susceptible to it, and particularly in skills we do not master well. A large majority of drivers (above 90%) believe that they are better than the average driver [44]. The Dunning-Kruger experiments indicate that the less we excel in something, the more confident "experts" we tend to believe we are [45]. Training and education might however help, as the experiments indicate that high competence reduces neglect and overconfidence.

The optimism bias may account for the frequent lack of realism in project planning and budgeting, in addition to other more tactical causes like securing the approval of a project by promising too much. They provide good arguments for applying quantitative project risk analysis, an application of QRA to projects. Realistic means for time and money should be calculated before a project is launched, with defined confidence limits. The assurance arrangement for public projects above 750 million NOK issued by the Norwegian Finance Department around the turn of the last century [46] is an example.

4.3 The confirmation bias and selection

People tend to come up with a hypothesis and then to find support for it, instead of trying to prove it wrong. When we are sufficiently confident about our presumption through confirming, we stop searching even though there could be better alternatives. The presumptions can be beliefs and rules for conduct of the form "if you do this under those conditions, the result will be such and such." Or they can be of simpler form, like stereotypes or weak correlations. Rules do not always produce the expected result. In behavioral terms, they are maintained on an intermittent or partial reinforcement schedule.

The confirmation bias may lead to the following of premature and false rules. Confirmation behavior also has positive effects as it serves to stabilize conduct and makes us more predictable, enhancing social acceptability. This is good if the rules are good. The downside is that one also becomes a reliable follower of rules that are not optimal and sometimes disadvantageous. Conformity and lack of innovation may be a high price to pay for social and political acceptance.

The Behavioral Insights Team (BIT), partly owned by the UK government, has identified the confirmation bias as one of the most prominent barriers against learning new skills and innovation [47]. In a recent study, the news consumption pattern of 376 million Facebook users was analyzed, showing that most users preferred to get their news from a small number of sources they already agreed with, further bolstering existing beliefs and preventing new insights [48].

4.4 Cognitive dissonance contingencies

Cognitive dissonance was introduced as an explanation of choice behavior in a situation of ambiguity [49] and is an example of an activity where both respondent and operant behaviors are present. The influence of affect in decision-making can be very significant [50]. One example of a bias is the "halo effect" that may occur when strong positive reactions are reflected over (conditioned) to otherwise neutral stimuli, an effect of stimulus generalization in classical conditioning. The opposite reaction is called the "horn effect." One aspect of emotional reactions is that they tend to be either positive or negative, experienced as pleasure or pain, broadly stated. When there is correspondence between, e.g., our belief and what we perceive, i.e., confirmation, there is a feeling of pleasure. When there is a discrepancy of some sort, the feeling is aversive.

In behavioral terms, and somewhat simplified, we might say that cognitive consonance, i.e., confirmation, produces behavior that is positively reinforced, while dissonance produces behavior that is under aversive control, resulting in either escape or avoidance behavior. Repeated instances of a response that effectively terminates a dissonance may become an automated avoidance response. A similar argument can be stated regarding consonance; it keeps behavior on a steady course. Automated behavior escapes our attention, and we do not notice neither the dissonance nor the consonance. Automated decision behavior on "autopilot" is energy efficient as deliberate considerations are not carried out. The result might be a failure to notice changes that should have induced another decision outcome.

If strong conflicting emotions and values are involved, cognitive dissonance behavior is difficult to modify [51], in particular if there are automated reactions involved. Accusations of being biased may strengthen the aversive emotion, as it moves the attention from the outcome of the decision to the person making the decision. Attribution to a person often leads to reactions of defense in the form of self-justification, and the behavior might become extremely resistant to change [52]. The defense behaviors have become automated avoidance behavior, either as self-justification, as counterattack of some form, or a combination of the two. Automated behavior is not reflected upon; it is subconscious. Automated self-justification includes self-illusions or blind spots for everybody to notice except the self-justifier himself [53], a favorite subject in many comedies.

To favor beliefs that are not falsifiable is a powerful form of defense against cognitive dissonance as it will protect against being proven wrong, which is an unpleasant experience for most people. Confirmation behavior may be maintained by both negative and positive reinforcements. An individual might have a confirmation behavior that is simultaneously maintained as avoidance/escape behavior and as positively reinforced gratification-seeking behavior (ref. the Pollyanna principle). Multiple contingencies might explain why confirmation behavior is a very strong default option and why falsification is so rare as a belief-testing strategy in daily life as well as in making risk management decisions.

4.5 The power of inadequate rules

Behavior maintained on a thin reinforcement schedule is more resistant to extinction and change than behavior that has been reinforced according to a continuous schedule [54], a phenomenon called the partial reinforcement extinction effect.

Rare confirmation of rules or beliefs implies that the following of such rules is maintained on an intermittent or partial reinforcement schedule. The result is that vaguely formulated rules and beliefs as are typical for qualitative risk analyses often are more resistant to change than rules that are more precise and correctly formulated. Unclear and rarely confirmed beliefs tend to have more dedicated and convinced followers than rules and beliefs that reflect reality more precisely. It is a paradox that, at least within certain limits, the less correct a belief is, the more convinced the believers are.

A similar phenomenon is observed in the Dunning-Kruger effect commented earlier, if less skilled implies have beliefs that are less correct and adequate than more skilled individuals. The Dunning-Kruger effect states that low-competence individuals tend to believe that their ability is higher than it really is. High-competence individuals have a more realistic view and may even slightly underestimate their performance. The authors comment that when people are incompetent in their strategies to achieve success, they suffer a dual burden: not only do they reach erroneous conclusions and make unfortunate choices, but also their incompetence robs them of the ability to realize it. Instead, they are left with the mistaken impression that they are doing just fine. This is also an example of the blind spot bias. As it is said: "First rule of the Dunning-Kruger Club: You do not know you're in the Dunning-Kruger Club." This statement is unfortunately valid for most biases, as being aware of a bias does not protect you from being biased.

5. Conclusion

In this chapter, a short history of risk management was presented before the most prominent cognitive biases were discussed. Due to their evolutionary past, they are natural to our behavior repertoire and difficult to change and avoid. To make choices under uncertainty constitutes an error-prone situation typical for risk management decisions that are influenced by both our phylogenetic and ontogenetic histories. The two learning processes have hugely different timescales and mechanisms of fitness for variation and selection. Comparing the two is like relating the elephant with the man sitting on the back of the elephant. They are both better off if they cooperate, in other words: how shall we minimize the negative effects of cognitive biases and how to utilize the positive effects of cognitions, i.e., thinking and verbal behavior, man's most precious virtues? The evolution of language and thinking gave man a crystal ball enabling imagination of a future that also contain age-related sickness, decline of physical and mental abilities, and inevitable death. Without the optimism bias, the evolution of mankind might have stopped when humanoids reached the stage of language and abstract thinking. Overconfidence is essential for innovation as it induces variation for selection to work on. Some of it ends in budget overruns, delays, and products that never make it to the market. The crucial question is to keep the "good" and avoid the "bad" variation and selection. How to balance these must be situation specific as there is a large difference between risk management in aviation and risk management in the development of digital consumer products. Flight safety leaves little room for variation, while digital gadgets must get to the market first with the new innovative product. Then, it is ok to fail given that you now and then hit a blockbuster. The market does the selection; the employees and management must secure sufficient variation, biased or not.

Confirmation behavior usually serves us well as it stabilizes conduct and makes it easier for others to predict our behavior, which is beneficial for building social relations. It also boosts self-confidence, because we perceive ourselves as more consistent and coherent than we actually are. Repeated often enough, the confirmation behavior can be automated, making us unaware of it. And when our behavior for whatever reason becomes inconsistent, as viewed by other people or by ourselves, the dissonance leaves us with an unpleasant feeling we seek to escape. We usually succeed, due to a well-equipped escape and avoidance behavior repertoire, developed over many years of our upbringing.

If the confirmation or avoidance behavior is maintained on a partial reinforcement schedule, it can become very resistant to change. The thinner the schedule, the more resistant the behavior is likely to become; within certain limits. Beliefs or rules that are less correct are confirmed more seldom than more correct beliefs. This opens for a subconscious and callous effect; the more wrong you are, the stronger you believe you are correct. In the discourse of science, falsification has therefore been proposed as the preferred scientific method for verification rather than confirmation [55].

Cognitive biases are human legacies from our behavioral past that may strongly influence decision-making in risk management. The research on these items is only in its infancy. That should however not prevent us from considering what we can do today, although we should be modest in what it is possible to achieve. The resistance to change and stealth like character of cognitive biases make them almost impossible to avoid. It is however possible to reduce their effect to some degree by reinstating QRA as the basis for risk management, extended with a review of relevant cognitive biases. The not-so-surprising solution is to let only well-qualified people take management positions. This might constitute a challenge as there normally are many to choose from. The best qualified are not necessarily those with highest self-confidence. It is rather the opposite, as the Dunning-Kruger experiments indicate. The less skilled people tend to unknowingly exaggerate their abilities. They are probably not the people you would like to make crucial decisions regarding risk, but you are likely to find them overrepresented among wannabe managers.

Regardless of industry, effective risk management implies that regulations must be in place that require QRA of high quality, and if risk levels are not acceptable, the measures that are most cost-effective must be identified and implemented, until the risk is within acceptable limits. This is the essence of a risk management system. If any of the steps are missing or carried out without the proper knowledge, compensation to those who suffered from the accident can be enforced. Using lawyers to promote safety by making the responsible pay compensation for damage is an example of corrective action; it is reactive as it is initiated after the accident has occurred. Accident prevention is however a much better strategy than damage compensation. A proactive way to promote safety is to set a much higher value on life. This policy should be made external to the involved stakeholders, e.g., by an official, public institution like NHTSA for road safety. A still better solution is to raise the decision of value of life above the different industry branches, as there is no good reason why the value of life should depend on what kind of work you do.

Acknowledgements

The author thanks Gunnar Ree for his helpful comments.

Conflict of interest

The author declares that he has no conflict of interests.

Author details

Jan Folkmann Wright
Department of Behavior Sciences, Oslo Metropolitan University, Norway

*Address all correspondence to: jan.wright@oslomet.no

References

[1] Bernsteain PL. Against the Gods: The Remarkable Story of Risk. New York: Wiley; 1996. ISBN: 13:9780471295631

[2] Aven T, Renn O. On risk defined as an event where the outcome is uncertain. Journal of Risk Research. 2009;**12**(1):1-11. DOI: 10.1080/13669870802488883

[3] Tversky A, Kahneman D. Judgment under uncertainty: Heuristics and biases. Science. 1974;**185**(4157):1124-1131. DOI: 10.1126/science.185.4157.1124

[4] Kahneman D. Thinking, Fast and Slow. Straus and Giroux, NY: Farbar; 2011

[5] Thaler RH. Misbehaving: The Making of Behavioral Economics. New York: W.W. Norton & Company; 2015. ISBN: 978-0-393-08094-0.

[6] Available from: https://www.iso.org/obp/ui/#iso:std:iso:31000:ed-2:v1:en

[7] Available from: https://www.coso.org/Pages/default.aspx

[8] ARMS Working Group. The ARMS Methodology for Operational Risk Assessment in Aviation Organisations; 2010. Available from: http://www.easa.europa.eu/essi/documents/Methodology.pdf

[9] Chouhan TR. The unfolding of bhopal disaster. Journal of Loss Prevention in the Process Industry. 2006;**18**(4-6):205-208. DOI: 10.1016/j.jlp.2005.07.025

[10] Vaughan D. The Challenger Launch Decision: Risky Technology, Culture and Deviance at NASA. Chicago: University of Chicago Press; 1996

[11] mv Herald of Free Enterprise: Report of Court No. 8074 Formal Investigation, Crown Department of Transport; 1987. ISBN: 0-11-550828-7

[12] McDonald LG, Robinson P. A Colossal Failure of Common Sense. The Inside Story of the Collapse of Lehman Brothers. New York: Random House; 2009. ISBN: 978-0-307-58833-3

[13] Orrell D. The Future of Everything: The Science of Prediction. First Thunder's Mouth Press Edition. New York: Avalon Publishing Group; 2007

[14] Baum W. Understanding Behaviorism. Behavior, Culture, and Evolution. 3rd ed. Hoboken: Wiley Blackwell; 2017. ISBN: 9781119143659 (pdf)

[15] Chapman AL. The anatomy of an accident. Public Health Reports. 1960;**75**(7):630-632

[16] Barlow RE, Proschan F. Mathematical Theory of Reliability. New York: John Wiley and Sons; 1965. Reprinted (1996) SIAM, Philadelphia, PA

[17] Rasmussen et al. Reactor Safety Study. An Assessment of Accident Risks in U. S. Commercial Nuclear Power Plants. NUREG-75/014; 1975

[18] WASH-1400 The Reactor Safety Study. The Introduction of Risk Assessment to the Regulation of Nuclear Reactors. Prepared by Reynold Bartel. NUREG/KM-0010; 2016

[19] Swain AD. A Method for Performing a Human Factors Reliability Analysis. Albuquerque, N.M.: Sandia Corporation Monograph SCR-685; 1963

[20] Swain AD, Guttmann HE. Handbook of Human-Reliability Analysis with Emphasis on Nuclear Power Plant Applications. United States; 1982. DOI: 10.2172/5752058

[21] Embrey DE, Lucas DA. Human reliability assessment and probabilistic risk assessment. In: Colombari V, editor. Reliability Data Collection and Use in Risk and Availability Assessment. Berlin: Springer; 1989. DOI: 10.1007/978-3-642-83721-0_27.

[22] Wright JF. Quantification of human error: The HUREL model. In: Proceedings of the Nordic Conference for Accident Investigators. Wadahl, Norway; 1985

[23] The Offshore Installations (Safety Case) Regulations; Statutory Instruments: No. 2885; 1992

[24] Borges, V. New edition — OREDA 2015 handbook. DNVGL.com. Det Norske Veritas and Germanischer Lloyd; 2015. Available from: https://en.wikipedia.org/wiki/OREDA

[25] Available from: http://www.humanreliability.com/

[26] Thaler RH, Sunstein CR. Nudge: Improved Decisions About Health, Wealth and Happiness. London: Penguin Books; 2009

[27] Angner E. A Course in Behavioral Economics. Palgrave MacMillan; Basingstoke, United Kingdom, 2012. p. 6

[28] Simon HA. Rational choice and the structure of the environment. Psychological Review. 1956;**63**(2): 129-138. DOI: 10.1037/h0042769

[29] Kahneman D, Tversky A. Prospect theory: An analysis of decision under risk. Econometrica. 1979;**47**(2):263. DOI: 10.2307/1914185. ISSN: 0012-9682

[30] Gigerenzer G, Goldstein DG. Reasoning the fast and frugal way: Models of bounded rationality. Psychological Review. 1996;**103**(4):650-669. DOI: 10.1037/0033-295X.103.4.650

[31] Perrow C. Normal Accidents. Living with High-Risk Technologies. New York, NY: Basic Books; 1984

[32] Weick KE, Sutcliffe KM. Managing the Unexpected: Resilient Performance in an Age of Uncertainty. San Francisco: Jossey-Bass; 2007

[33] Reason JT. Managing the Risks of Organizational Accidents. Aldershot, England: Ashgate Publishing Limited; 1997

[34] Available from: https://www.motherjones.com/politics/1977/09/pinto-madness/

[35] Birsch D, Fielder JH. The Ford Pinto Case: A Study in Applied Ethics, Business, and Technology. Albany, NY: State University of New York Press; 1994

[36] Bratland D. Statistics. Available from: https://commons.wikimedia.org/wiki/File:US_traffic_deaths_per_VMT,_VMT,_per_capita,_and_total_annual_deaths.png

[37] Kagan RA, Scholz JT. The criminology of the corporation and regulatory enforcement strategies. In: Hawkins KO, Thomas JM, editors. Enforcing Regulation. Springer, Kluwer-Nijhoff; 1984. DOI: 10.1007/978-94-017-5297-8_4

[38] Wright JF. Risk management: A behavioral perspective. Journal of Risk Research. 2018;**21**:710-724. DOI: 10.1080/13669877.2016.1235605

[39] McKay RT, Dennett DC. The evolution of misbelief. Behavioral and Brain Sciences. 2009;**32**:493-561. DOI: 10.1017/S0140525X09990975

[40] Puri M, Robinson DT. Optimism and economic choice. Journal of Financial Economics. 2007;**86**:71-99. DOI: 10.1016/j.jfineco.2006.09.003

[41] Seligman MEP. Helplessness: On Depression, Development, and Death.

San Francisco: W.H. Freeman; 1975. ISBN 0-7167-2328-X

[42] Moore DA, Healy PJ. The trouble with overconfidence. Psychological Review. 2008;**115**(2):502-517. DOI: 10.1037/0033-295X.115.2.502

[43] Sharot T. The Optimism Bias: Why We're Wired to Look on the Bright Side. London: Little, Brown Book Group. Kindle Edition; 2012. ISBN: 978-1-78033-263-5

[44] Svenson O. Are we less risky and more skillful than our fellow drivers? Acta Psychologica. 1981;**47**:143-151

[45] Kruger J, Dunning D. Unskilled and unaware of it: How difficulties in recognizing one's own incompetence lead to inflated self-assessments. Journal of Personality and Social Psychology. 1999;**77**(6):1121-1134. DOI: 10.1037/0022-3514.77.6.1121

[46] In Norwegian. Available from: https://www.regjeringen.no/no/tema/okonomi-og-budsjett/statlig-okonomistyring/ekstern-kvalitetssikring2/hva-er-ks-ordningen/id2523897/

[47] Available from: https://www.behaviouralinsights.co.uk/reports/how-confirmation-bias-stops-us-solving-problems/

[48] Schmidt AL, Zollo F, del Vicarioa M, Bessi A, Scala A, Caldarelli G, Stanley HE, Quattrociocchi W. Anatomy of News Consumption on Facebook. 2017. Available from: www.pnas.org/cgi/doi/10.1073/pnas.1617052114

[49] estinger L. A Theory of Cognitive Dissonance. Reissue. ed. Stanford, California: Stanford University Press; 1957

[50] Finucane ML, Alhakami A, Slovic P, Johnson SM. The affect heuristic in judgments of risks and benefits. Journal of Behavioral Decision Making. 2000;**13**:1-17

[51] Lerman DC, Iawata BA. Developing a technology for the use of operant extinction in clinical settings: An examination of basic and applied research. Journal of Applied Behavior Analysis. 1996;**29**(3):345-385. DOI: 10.1901/jaba.1996.29-345

[52] Tavris C, Aronson E. Mistakes Were Made (but not by me): Why We Justify Foolish Beliefs, Bad Decisions, and Hurtful Acts. Orlando, Florida: Harcourt; 2007

[53] Pronin E, Lin DY, Ross L. The bias blind spot: Perceptions of bias in self versus others. Personality and Social Psychology Bulletin. 2002;**28**(3):369-381. DOI: 10.1177/0146167202286008

[54] Rescorla RA. Within-subject partial reinforcement extinction effect in autoshaping. The Quarterly Journal of Experimental Psychology. 1999;**52B**(1):75-87

[55] Popper K. The Logic of Scientific Discovery. Abingdon-on-Thames: Routledge; 2002 [1959, 1934]. p. 66. ISBN 0-41527843-0

Chapter 3

Functional and Technical Methods of Information and Risk Communication

Carine J. Yi and Tim Park

Abstract

Risks of natural and anthropogenic disasters can appear at any moment without warning. All levels of government agencies from federal to township, and the systems of the specialized agencies, like weather-specialists, flood control, electricity suppliers, and educational organizations are standardized and the communication has been improving. The systems of the governments and agencies, meteorologists, flood control, electricity suppliers, and educational organizations are standardized and the communication has been improving. In this chapter, government manuals, procedures for agencies and professional responders, and public awareness, perceptions, and capabilities, are reviewed in three international cities: Seoul, Tokyo, and Toronto. Each city is unique with experiences of different disasters. Communication of supports and vital information of risks without understanding the language and culture of the people may lead the public to large-scale panic. Individuals can access government websites and interpret the information, like WebGIS maps, and risks by themselves. In terms of risk communication, all urbanized cities require their own specialized risk management with reasonably effective technologies, which enhance community resilience. Even better is to have a measure of development for the care of the public after the disaster to help the people get back on their feet, such as various public insurances.

Keywords: disaster risk management, risk communication, WebGIS, SNS, Seoul, Tokyo, Toronto

1. Introduction

1.1 Risk on disasters

The United Nations [1] summarized the cities' risk of exposure to natural disasters. Of the 1692 cities with at least 300,000 inhabitants in 2014, 944 (56%) were at high risk of exposure to at least one of six types of natural disaster (cyclones, floods, droughts, earthquakes, landslides, and volcano eruptions), based on the evidence on the occurrence of natural disasters over the late twentieth century. Floods were the most common type of natural disaster affecting cities, followed by droughts and cyclones. Twenty-seven highly developed cities, such as Seoul and Tokyo, have to deal with higher risk of exposures to three or more types of disasters and are more vulnerable to disaster-related economic losses and mortality.

1.2 Risk communication in emergency situation

According to Ortwin Renn [2, 3], the field of risk communication was initially developed as a means of investigating how best expert assessments could be communicated to the public, so that the tension between public perceptions and expert judgment could be bridged. Risk communication links scientists, decision makers, professional performers, communication specialist, and the public. Also, risk communication is to assist stakeholders and the public at large in understanding the rationale of a risk-based decision, and to arrive at a balanced judgment that reflects the factual evidence about the matter at hand in relation to their own interests and values [4].

These days, advanced technologies are applied for better effective communication on answering the "want to know" questions that Michael S. Baram [5] summarized the concept of "genetic right-to-know laws" on hazard communication in manufacturing work environment. Risk communication between governmental agents and individual personnel would be focused in different ways. In most of the cases under the emergency circumstances, any government agents have to focus on their urgent roles rather than to contact with civilian individuals. Conversely, any individual personnel expect that they can receive the information that they want to know or need to know. In terms of the interactive risk communication between groups, such as different government agencies, agents, and individuals are often appearing to lack in interactive communication because of the differences of their own interests. Each level of government agencies including state government, municipality, township, and even specialized agencies such as weather-related groups, flood control, electricity suppliers, and educational organizations can send out the general information with professionally analyzed results.

Today, all types of information are shared via the Internet, with which anyone can freely access details about others' daily lives. The information can contain disasters' natural characteristics and mechanisms with forecasts, damage level and its costs, information acquisition methods, evacuation routes, location of shelters and emergency supplies, school re-openings, hospital capabilities and capacities, information on relocation of patients, road conditions, information-related insurance, and even family member's safety confirmation. Among communication devices, including landline phones and computers, mobile phones are the most effective and primary devices for global communication [6]. In the case of the 2011 Great East Japan Earthquake and subsequent tsunami, some victims, especially isolated victims, had no connection to the outside and experienced great fear and anxiety on obtaining information they want; the transmission towers of mobile phone were broken and they lost electricity and thus lost the ability to even recharge their mobile phones [7]. For risk communication under severe circumstances—for example, flooding, earthquake, wildfires, terror, explosions, and others—information is the most important matter for everyone.

Information is available from multiple sources, from online sources and web-based media, personal conversation, mass media, and official and unofficial reports. However, individuals uniquely internalize this overwhelming information and judgmentally amplify their risks in various ways and degrees [8]. Rumors, or gossip, pass around easily (and thus, fairly quick) which may lead people to wrong directions or decisions, especially under emergency circumstances. Hence, the accurate information and the correct and effective instructions to maintain the safety of people and the society are one of the important roles for the government agencies.

Although every government tries to develop a disaster risk preparedness manual and educate public to enhance the risk perception, disaster risk governance is fairly a new concept [8]. Risk governance requires more one-directional communication rather than interactive communication. Arnold Howitt emphasizes the crisis management communication, which interprets risk more narrowly with four fundamental rules that apply to the government agencies (as cited in [9]);

- Say What You Know
- Say What You Are Doing
- Say What Others Should Do
- Offer Perspective

The information must contain two different phases at least; phase of preparedness prior to an emergency and phase of the said emergency. The preparedness phase will suggest how to get ready for risks, while the latter phase should give an idea to the public where the safe place is located or how to protect themselves when the disaster occurs.

1.3 Utilize technology for risk communication

During the time of crisis or emergency, the critical issue is whether the people understand what to do. To empower people with knowledge and information before crisis (the "nonemergency time" or "disaster preparedness phase"), government agencies use websites to provide access to this information. Mapping technologies, such as geographic Information System (GIS) associated with Web application environment, are an effective method of disaster-risk communication to amplify the information to broader audience and population. When incidents or disasters occur, governmental agencies implement intensive tasks such as scientific analysis, forecasts, finding safe places and access to get there, dispatch rescue teams and give instructions for safety activities, and so on. Under these circumstances, government staffs frequently appear on the media to try to communicate with the public. In the circumstance under emergency, GIS technology has precisely functioned and clearly proved its effectiveness. 9.11 World Trade Center Attack was one of the significant incidents, and GIS technology was well-exercised. Since this event, GIS and Web technology was combined and developed for the next advanced world technological risk communication.

Social network services (known as SNS) such as Facebook, Twitter, and YouTube are the mostly used web tool for interactive communication in emergency situations. These service users are not only victims, but also supporters and governmental agencies that send the detail of information on how/when/where to get help [7]. The real-time situational information from the victims—for example, victims' conditions, actual detailed help list, health situation of the victims, security problems, and other sorts of information—is actively exchanged through these SNS services.

2. Risk management in three cities

The insurance company Lloyds pointed out at the report of Lloyd's City Risk Index 2015–2025 that mega-cities (such as Seoul, Tokyo, New York, Hong Kong, Shanghai, and London) are exposed to risks of combination of high economic value losses and high exposure to both natural catastrophes and manmade risks. Cities with high asset values are the most financially exposed in absolute terms [10].

The comparison of three large cities, Seoul, Tokyo, and Toronto, is very unique work because of their different profiles of risks they face in their respective countries. Their common characteristics are that all are very urbanized capital towns with the highest population in each county as shown in **Table 1**, the center of the financial activities, and very globalized. Tokyo, for example, is one of the largest cities in the world with the population and economy, and the capital city of Japan, which is a country well recognized as the disaster kingdom of the world. Toronto, Canada is a metropolitan city that is a popular choice for newcomers including immigrants and refugees, short-term residents, or visitors that are exposed to different categories of potential disasters and risks than Japan. On the other hand, Seoul, Korea is a dense mega-city that is close to Japan, China and also to North Korea. Risks for Seoul are different from the risks of Tokyo or Toronto.

According to the report of the London-based insurance company, Lloyds [10] classifies 18 types of threat risks with 5 categories: finance and trade (market crash, sovereign default, and oil price shock), geopolitics and society (terrorism), natural catastrophe and climate (earthquake, wind storm, tsunami, flood, volcanic eruption, drought, freeze, and heat wave), technology and space (nuclear accident, power outage, cyber-attack, and solar storm), and health and humanity (human pandemic and plant epidemic). They assess the threat of all 18 types and how likely that city can experience a number of representative scenarios of different magnitudes from that threat (3 representative scenarios). Lloyds developed metrics for economic consequences and applied the equation for the 'expected loss (loss × the probability)' which is called gross domestic product (GDP)@risk for each city. This model is implemented for identifying which cities and threats are the most important.

2.1 Seoul, Korea

2.1.1 Background

Seoul is the capital city of the Republic of Korea (South Korea), the place for a habitat of 10,178,395 people [11]. As of the end of 2010, the area of Seoul is 605.25 km^2, which is about 0.6% based on the total land area (100,033 km^2) at the same time [12]. The population density of Seoul has continuously increased with population growth. After the Korean War and temporary population drop, the city of Seoul accepted many people into its area, which followed population boom in the city and forced the architecture of Seoul upward to address the housing problems, building tall apartments in the 1990s to house the increasing population growth. As of the end of 2010, the population density of Seoul is 17.473 people/km^2 [12, 13].

City	Country or area	Statistical concepts	City population (thousands)			Average annual rate of change (percentage)		City population as a proportion of the country or area's total or urban population in 2016 (percentage)	
			2000	2016	2030	2000–2016	2016–2030	Total population	Urban population
Seoul	Republic of Korea	City proper	9878	9779	9960	−0.1	0.1	19.6	23.7
Tokyo	Japan	Metropolitan area	34,450	38,140	37,190	0.6	−0.2	30.1	32.1
Toronto	Canada	Metropolitan area	4607	6083	6957	1.7	1.0	16.8	20.5

Table 1.
Three cities' population (modified, [1]).

Korea is the only divided country in the world that has been in a state of continuing confrontation with North Korea since the end of the Korean War in 1953. Seoul is only 23 km (in the shortest distance) away from North Korea. The confrontation with North Korea for 65 years has kept Seoul citizens in a high level of crisis cognition. Nonetheless, with a remarkable economic growth, foreign residents are increasing since 2014. According to the Brookings Institute, Seoul-Incheon is placed fourth with total GDP with 846 billion US dollars of the world's largest city economies in 2014, which was almost the same as Los Angeles (860) (Tokyo was the first with GDP 1600) (as cited in [14]. In 2015, nominal GDP of Seoul-Incheon (population 25,095 thousands) was 25,095,903 million US dollars and nominal GDP/capita in 2015 was 466 36,002 million US dollars [15]. Seoul is one of the richest cities in the world.

2.1.2 Risks in Seoul

It seems that all these experiences influenced the surveys that are conducted by the Statistics Korea [16] every 2 years for the 10 social indicators system. This data will be available on May 16 for approximately 39,000 household members aged 13 years or older residing in the 25,843 sample households nationwide for the five indicators in family, education, health, safety, and environment. Regarding awareness of social safety, 20.5% of the respondents felt that the society was "safe" overall. The most significant anxieties about the social safety were crime (50.8%), traffic accidents (47.6%), new diseases (42.8%), and information security (42.5%), in relative order. For the question of emergency and emergency situations, most (96.8%) "knew the telephone number (119, etc.)" in the case of an emergency. Seoul residents feel that social indicators are more severe real-life threats than natural disasters. The result of the survey is summarized in **Table 2**.

Lloyds [10] defines that Seoul is facing the risk of windstorm (44.68 billion) economic loss as shown in **Figure 1** the most, and this is 3.5 times more than the oil price stock threat of 12.72, at the second place. Therefore, Seoul is very vulnerable to typhoons. In addition, due to the proximity of oil price shock and market crash, and because Korea has experienced it already in the past, Korea has to react sensitively to the international economic market (1997–1998 Korea financial crisis, which was triggered by 1997 Asian financial crisis). However, earthquake and tsunami were assessed as zero risk in Seoul, while they were the most threatening disaster risk for Tokyo. The total GDP@risk is estimated as103.50 billion US dollars with all 18 threats.

2.1.3 Improvement of crisis management and risk communication

For a long time, the constructed structure of the governance during an emergency is hierarchical. A national crisis incident that occurs within a ward is reported by the staffs of the ward government to the chief of the ward, which is then reported to the municipal council; the municipal council summarizes the issue(s) and, then, reports to the provincial board. The members of the provincial board collect and then report to the federal government, where the chief of that level (usually the president) is the decision-maker for an incident. The decision is then trickled down from the province, to the city, and down to the ward, for the emergency response to take place. This pyramidal communication structure is inherently slow and involves too many individuals in which a report must be filtered through. This introduces complexities such as missing information, distortion of crisis, misunderstanding of issues, and/or severity of the emergency.

		Total	Safe	Very	Relatively	Normal	Not safe	Relatively	Not at all
Overall social safety	2016	100.0	13.2	1.1	12.1	41.2	45.5	36.1	9.4
	2018	**100.0**	**20.5**	**1.3**	**19.2**	**48.2**	**31.3**	**27.6**	**3.7**
National security (war potential, North Korean nuclear issue)		100.0	31.1	3.2	27.9	35.6	33.3	27.3	6.0
Natural disasters (typhoons, floods, and earthquakes		100.0	22.8	1.6	21.1	39.6	37.6	32.2	5.4
Buildings and facilities (collapse and explosion)		100.0	23.9	2.1	21.8	43.3	32.8	28.0	4.7
Traffic accident		100.0	13.1	0.9	12.2	39.2	**47.6**	38.4	9.3
Fire (including forest fire)		100.0	20.9	1.8	19.2	48.0	31.1	26.6	4.5
Food (defective food and food poisoning)		100.0	25.4	2.3	23.2	43.4	31.1	25.5	5.7
Food security (grain flooding and food shortages)		100.0	37.8	5.9	31.9	43.7	18.5	15.4	3.1
Information security (computer virus and hacking)		100.0	17.5	1.6	15.9	40.0	**42.5**	33.4	9.0
New diseases (new virus)		100.0	16.7	1.4	15.3	40.5	**42.8**	34.7	8.1
Crime		100.0	17.2	1.8	15.4	32.0	**50.8**	38.0	12.8

Bold values indicate the most recently surveyed and the highest values in the categories.

Table 2.
Recognition of social safety (unit: %).

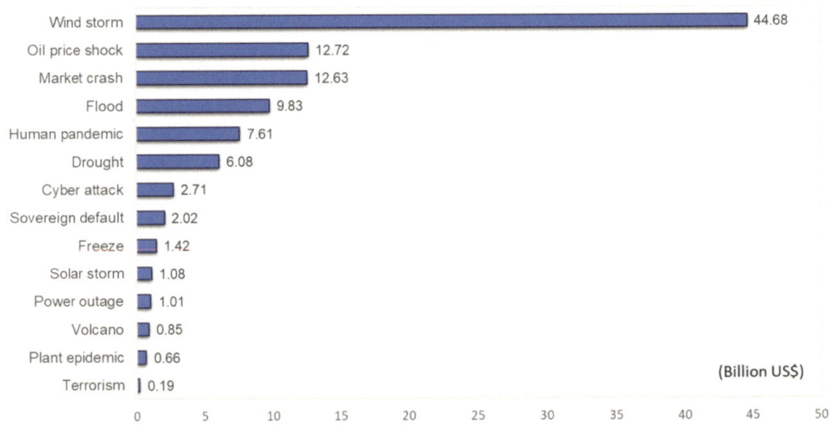

Figure 1.
GDP@risk metrics for Seoul (modified, [17]).

There was a significant incident, which was an important turning point and a trigger, which forced Korea to review and overhaul its crisis management system, as well as replace the government administration with the decision from presidential authority; the significant incident is known as the "Sinking of MV Sewol" in 2014. The South Korean ferry sank while carrying 476 people, mostly secondary school students from Danwon High School (Ansan City). In total, 304 passengers and crew members died in the disaster [18, 19]. Of the approximately 172 survivors, more than half were rescued by fishing boats and other commercial vessels that arrived at the scene approximately 40 minutes after the South Korean coast guard. As the Korean government's attitude toward disaster responses and all the procedures of rescue work were inadequate, regulatory authorities who directed and supervised them were socially criticized, and their attitude of avoiding their responsibility was enough to buy public sympathy [20, 21]. Perhaps, this incident is one of the most suitable examples of Heinrich's law[1].

With these experiences, with sufficient stimulation in the city of Seoul, the city of Seoul authorities made an overall review of the disaster response and initiated the reorganization. The city of Seoul developed the disaster management communication guideline in 2014 [22]. Newly developed "the City of Seoul Crisis Management Communication Guidelines" emphasize four priority values of risk communication capabilities: empathy, transparency, professionalism, and responsibility. **Figure 2** is the diagram that describes the concept of priority value of the risk communication.

In order to prevent the confusion of reporting system, the Seoul Metropolitan Facilities Management Corporation improved its reporting system in case of emergency situation as shown in **Figure 3**. The direction of the arrows indicates the direction of reporting of an incident; one-way arrows represent only reporting and lack of interactive communication (one states "confirm occasionally", which indicates that no actions are taken) and a head group that takes control or responsibility in implementing a plan during an emergency [23]. From this diagram, for example, the hydro-related disaster, the decision makers are far and dislocated from the incident scene, which means that the duration to report to the decision makers takes longer time and the order (decision to act) demands for more time to be communicated back to the scene for an actual work. Thus, there is no clear indication

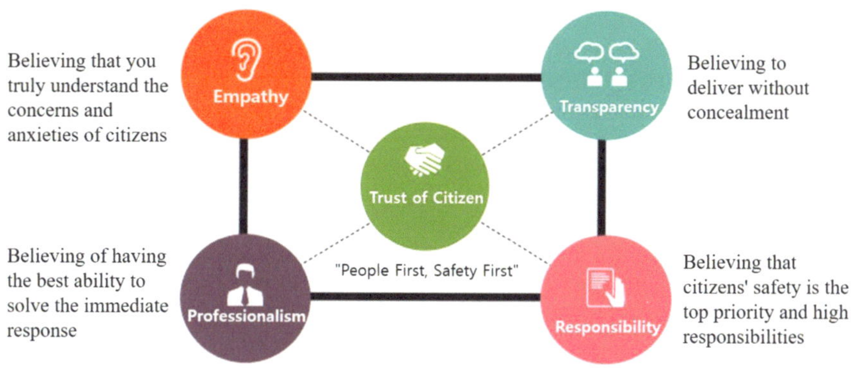

Figure 2.
Four priority values of risk communication capabilities and its expected value criteria (modified, [22]).

[1] Heinrich's Law: Heinrich's work is claimed as the basis for the theory of behavior-based safety by some experts of this field, which holds that as many as 95% of all workplace accidents are caused by unsafe acts. (Source: https://en.wikipedia.org/wiki/Herbert_William_Heinrich).

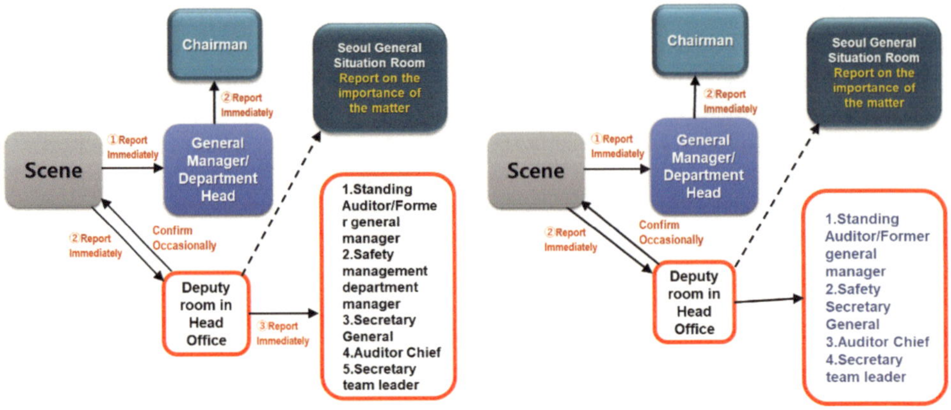

Figure 3.
Diagram of situation report flow between departments in government offices [23].

of how much the decision makers understand the real situation and take actions or responsibility. It is easy to assume that such structure would result in delayed communication and introduce unnecessary complications (especially in between the communicators), and overly long decision-making time. Meanwhile, it does not address who takes charge or takes responsibility to control or direct an emergency; thus, the field works are forced to deal with victims, the crisis, and their own safety on the line without a clear authority, permission, or instruction.

Realizing the shortcomings and inefficiencies of the hierarchical or pyramidal communication system, the national emergency management agency of Korea (normally known as the Department of Fire and Emergency) developed their own drill manual [24] with incorporating with other agencies. Based on convergence of administration, the mutually developed cooperation and integration of the agencies divide the complicated workflow.

Despite the efforts made by the Seoul Metropolitan Government to revise its crisis management guidelines after the 2014 MV Sewol sinking incident, the government's ability to manage crises has been criticized, again, widely and intensely on the way they responded to the 2015 MERS outbreak incident [9]. A joint panel of experts from the World Health Organization and South Korea announced that the South Korean government's failure to share information quickly with the public and establish an efficient disease-control system contributed to worsening the outbreak of Middle East respiratory syndrome (MERS), which has killed 14 people in the country [25] and this event was another significant incident to receive social criticism of the government and its stakeholders and their awareness of crisis management.

2.1.4 The IT country: utilizes advanced IT technology

As an IT powered country, the capital city of Seoul is working on risk communication by operating a well-organized portal-website as shown in **Figure 4** (top image). This website provides breaking news, action procedures, evacuation facilities, related public agencies, weather information, and the information for the press. Also, the website has a linkage option that jump to the new window where anyone can join the discussion with certain subject, leave comments, and apply suggestions. GIS interactive map is generated by clicking of the icon on the right panel with the information icon.

Although technology is shifting to Cloud, WebGIS technology is still used widely, as those who use the technology are divided in the proficiency of available technologies, but the trend of GIS is heading toward cloud-based technology. GIS

Figure 4.
The city of Seoul safety web portal, which is called "Seoul Safety Nuri" that gives information about current disaster status and suggests a direction where shelters are located and its capability [26].

maps allow users to identify certain facility locations with accurate coordination, query function to call up the associated database information, so end-user can get information by several times of clicking. The disaster maps of Seoul give an option to generate satellite image as background map (**Figure 4**: bottom left) or topography map (**Figure 4**: bottom right); then users can recognize where they are located, where they want to go, what to look for, or identify facilities and its basic information, (such as the name of facility, address, functions, and its capability). For example, the map of evacuation shelters provides information about shelters: earthquake indoor/outdoor shelters, extreme heat weather shelters, water facilities, storage area for snow removal equipment, and general disaster maps. This integrated GIS map has functions associated with the database, which are very limited and still need improvement for the real-time support system. Despite Seoul's copious efforts and technologies, this website has only provided information in Korean, and no other languages are available at this time.

Among the weather-related disasters, it is clear that the typhoon and flooding are big threats when considering the Han River (a large river that passes through the middle of Seoul from east to west). The history of Seoul's modern sewerage dates back to the opening of the late nineteenth century. At that time, the importance of the sewage treatment for the prevention of infectious diseases and the improvement of public sanitation was recognized, and drainage function was moved from the river to the modern sewer model [27].

As of the end of 2015, the city of Seoul is discharging sewage and rainwater to four water reclamation centers through the sewage lines (about 10,000 km), manholes (260,000), and rainwater catchers (470,000). The amount of sewage that has been treated and flowing into the rivers reaches 4.98 million cubic meters/day. Currently, the city of Seoul has faced a new era of change [27]. As global climate change intensified, Korea is entering change from a sub-tropical climate of four seasons in temperature to a more two seasons of extreme temperature differences. Seoul faces a large risk of natural disasters such as urban flooding and drought. As the phenomenon of heat island occurs more frequently, more powerfully, and for longer periods, it is necessary to cope with thorough water management.

2.2 Tokyo, Japan

2.2.1 Background

As of 2014, the greater Tokyo area was ranked as the most populous metropolitan area in the world (**Table 1**; [1]). The 23 Special Wards of Tokyo were formerly Tokyo City. Tokyo is the one of the largest cities in the world with an estimated population of 13.491 million (as of October 1, 2015); in 2191 km^2, the area of Tokyo is 0.6% of the total area of Japan. Most of the population reside in 23 wards in where 9241 million, the Tama area, 4.223 million, and the Islands, 26,000. With a population density of 6158 persons/km^2, Tokyo is the most densely populated prefecture in Japan. Tokyo has 6.946 million households, with an average of 1.94 persons/household. The number of foreign residents according to the basic resident register is 440,000 as of October 1, 2015 [28].

The general Tokyo budget of the 2016 fiscal year was about 7 trillion yen, and a total amount of about 13 trillion yen will be reached by adding public enterprise accounts as special budget, such as water supply, sewerage, and subway. It is comparable to the national budget of Sweden and Indonesia where it would differ slightly due to the exchange rate [29].

2.2.2 Disaster kingdom of Japan: historical natural disasters leave lessons

Tokyo is the capital city of Japan, the kingdom of disasters. In accordance with the Lloyd's report, Tokyo is ranked second for volcanic eruption risk, third for heat wave, fourth for windstorms, and sixth for earthquakes. Mega city, Tokyo, is highly exposed to man-made and natural threats, which makes it the world's second-riskiest city to live in after Taipei; Tokyo risks a possible economic loss of $153.28 billion of its GDP by the likelihood of events occurring over the next 10 years. Exposure to windstorms may bring Tokyo a loss of $29.06 billion, while exposure to earthquakes and flooding may trigger a loss of $18.83 billion and $17.65 billion, respectively. However, London-based Lloyds said that Tokyo's exposure to potential loss is comparatively low and accounts for only 10.44% of its $1.47 trillion in estimated annual GDP, making it "one of the richest cities." [1, 10].

Lessons learned from disasters often come after catastrophic economic losses [30]. Andrew Coburn [17] summarized the economic loss after Great Hanshin earthquake January 17, 1995, Magnitude 7.3 in a slide at the 2015 seminar in London, UK. As many of the problems posed by this earthquake were highlighted, Japan has renewed many policies, and the citizens' awareness of the natural disaster has increased. After the 2011 Japan earthquake and tsunami, the natural disaster caused secondary disasters such as explosions and radioactive spills of nuclear power plants, and the fear of a direct underground earthquake in the metropolis of Tokyo was heightened.

Japanese society is a well-organized one. Particularly, for Tokyo, rail is extremely important as a means of transportation for people and goods, and is well-organized and maintained diligently considering that Tokyo has many commuters from suburban area. The national census in 2010 lists the daytime population of Tokyo as 15.576 million people, which is 2.417 million (or 1.2 times) more than the nighttime population figure of 13.159 million. The daytime population index is 118.4 against the nighttime population taken at 100. This difference is caused by the population of commuting workers and students, constituting a daytime influx from mainly the three neighboring prefectures of Saitama, Chiba, and Kanagawa [28].

During the 2011 Japan Earthquake and tsunami, a large population of commuters who has difficulty returning home were stuck in Tokyo until the public transportation reopened, and some of them chose to walk because they could not find the accommodation. According to estimates based on the Internet survey published by the Cabinet Office of Japanese Government on 22 November 2011, due to the 2011 Japan Earthquake and tsunami, approximately 3.52 million people in Tokyo, about 670,000 people in Kanagawa Prefecture, about 520,000 people in Chiba Prefecture, about 330,000 people in Saitama Prefecture, and about 100,000 people in the southern part in Ibaraki Prefecture, total 5.55 million people in the metropolitan area, experienced difficulty returning home or were unable to return home. Hiroi et al. [31, 32] conducted the internet survey 2 weeks after (March 25–28, 2011) the 2011 Japan Earthquake and tsunami that occurred on March 11, 2011. According to this survey, about 28% of the outsiders, when the earthquake occurred, could not return home during the day. Also, as for the question of what was the most troubled matter by this earthquake, 71.1% responded disconnected mobile phone system/reception, followed by decommissioned public transport such as rail (46.4%), inability to contact the family (37.5%), the home telephone not working (35.2%), and short message by the cellular mail not working (32.9%). Wendling et al. [33] pointed out that under certain unusual circumstances, such as natural disasters, victims are desperate to obtain reliable and relevant information from any type of media. Social media can play an important back-up role in dissemination warning and response information if traditional services are overwhelmed by demand.

The idea of risk factor as a perceived magnitude of a disaster does not play a very important role in the risk perception of natural hazards [8]. However, in terms of heuristic rule for Japanese people at least, it is hard to agree with such statement. Japanese are being educated about the natural disaster risks and have been trained on how to protect themselves and the way to follow the evacuation procedures. Most of the populations start from young age of kindergarten. Understanding the risk is critical for Japanese people, and therefore, the Japanese government prepares manuals and evacuation rules efficiently.

2.2.3 Learn together and survive together

The Tokyo utilizes IT technology as well. Tokyo developed a special website, Disaster Prevention Website [34], which provides preparedness information and guidelines to follow on emergency circumstance. Much of the information provided with the text-to-speech functions for those with vision impairment. However, such features must be manually activated via clicking an icon. Thus, the efficiency of the use of such function is not yet verified. The organization tree appears to be strict and usually follows a top-down command structure. The crisis management task team is created to work on establishing a crisis management system to gather and analyze information and prepare strategic measures, while disaster prevention system describes its liaison relationship with the national government, municipalities, and other agencies based on information from the Disaster

Prevention Center as depicted in **Figure 1**. However, this structure only explains the relationship between governmental agencies. This supposed relationship should not only explain the relationship between the provision of information and responsibilities of the intergovernmental departments, but should also explain the sharing of information and communication flow between the government and the residents—which **Figure 5** fails in doing. As the risk communication is a multi-way interactive communication between sectors, this structure can lead to stagnation of communication among the end-users, especially those who are the most vulnerable to disasters.

The Tokyo authorities have compiled a manual called "Disaster Preparedness Tokyo (東京防災)" to help households get fully prepared for an earthquake directly hitting Tokyo and other various disasters (**Figure 6**).

"Disaster Preparedness Tokyo" is tailored to the various local features of Tokyo, its urban structures, and the lifestyles of its residents and contains easy-to-understand information on how to prepare for and respond to a disaster. This information will be useful now and in the event of an emergency. This manual consists of introduction, five chapters, and a manga comic, which provide guidance from the preparedness actions, survival tips, disaster facts and acquisition of information, to evacuation procedures. It is unique in that it uses pictures to explain to the foreigners who do not know Japanese or English. In particular, comics explain how to act without falling into panic by briefly showing how to deal with an earthquake. In this manual, Tokyo authorities have a lot of explanations on how to deal with an earthquake. Also, this means how frequently Tokyo is experiencing earthquakes and trying to reduce earthquake damages and human losses. In the mega city of Tokyo, which has a huge population and economic power, earthquakes are the most threatening risk.

2.2.4 Maps are ready for you

The disaster prevention homepage (**Figure 7**) provides a lot of information as much as they can open to the public. The map utilizes the WebGIS function, and it is well used to upload the prepared disaster-related information such as the location of the facility and road construction information. However, these maps, although

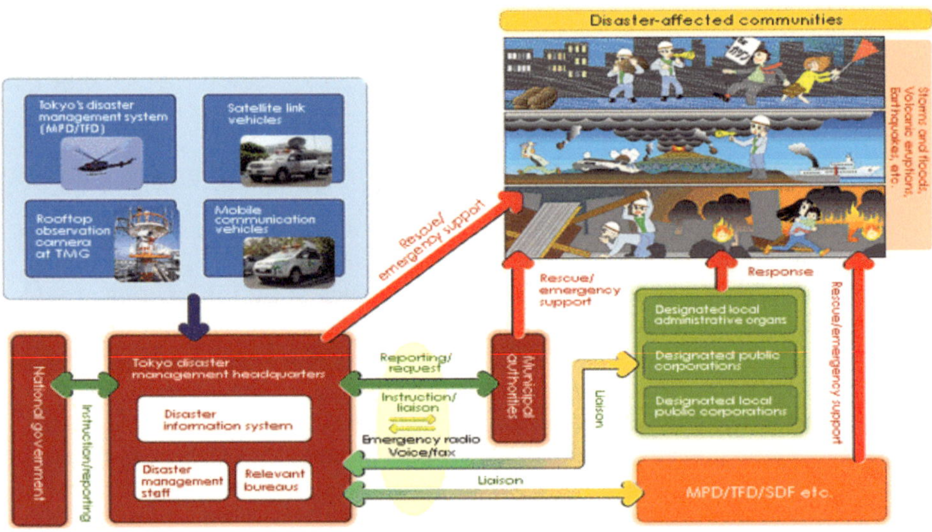

Figure 5.
Workflow of the disaster prevention system.

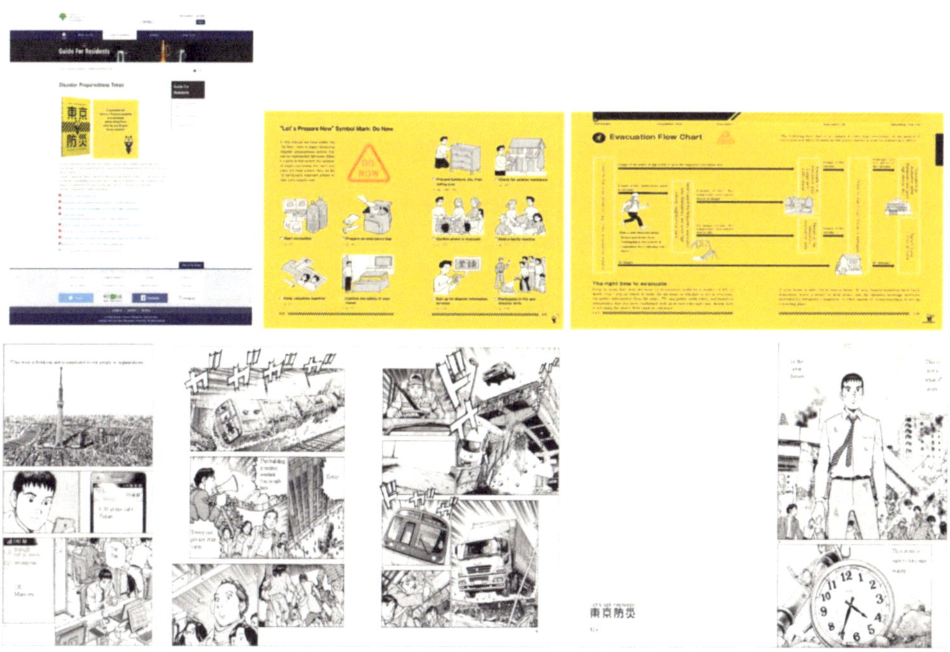

Figure 6.
The English website of "Disaster Preparedness Tokyo" manual; an example of chapters and parts of manga comic.

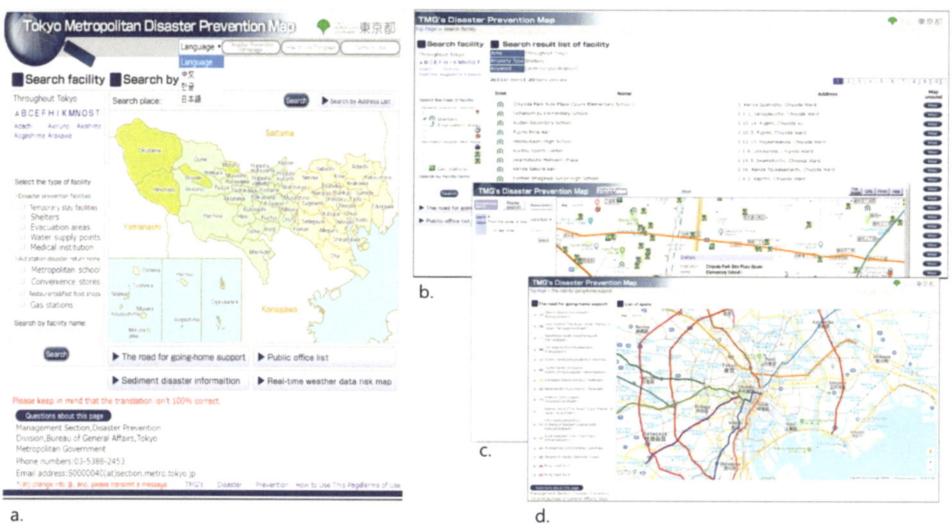

Figure 7.
Tokyo provides disaster prevention map, which shows the list of the facilities and locations by different disaster types. (a) Interface of Tokyo metropolitan Government's disaster prevention map with selection functions, languages, facilities, and related emergency information according to the name of the administrative district. (b) Listed queried information. (c) Google Maps with facility location and its information retrieved by clicking the map icon. (d) The road going-home support function shows where to select a road name and it will display the main crossing points in the road by a list and icons.

used extensively during peacetime, are not guaranteed to be used during the time of disaster crisis to provide real-time updates of the emergencies. The maps focus on the Tokyo metropolitan, which includes 23 wards and neighboring cities under Tokyo metropolitan address. Interface of the map has several functions: languages (Korean, Japanese, English, and Chinese), facilities (temporary stay facilities, shelters, evacuation areas, water supply points, and medical institutions) and facilities that are run by organizations and private owners that provide current-state disaster status (metropolitan schools, convenience stores, restaurants and fast food shops, and gas stations). It seems that these functions are selected by learning from the 2011 Japan Earthquake and tsunami (**Figure 7a**). There are four additional support options: road information supporting returning home, public office list, landslide disaster-related information, and real-time weather data risk map, which is linked to the Japan Meteorological Agency to obtain the weather data. Users can review the desired facility or function by clicking the check box and the query function used to open a list with basic information about related facilities (**Figure 7b**). Then, by clicking on the map icon, the Google Maps will retrieve Information about the selected facility and display in the form of a memo next to the icon of the facility, so the users can quickly find out relevant information (**Figure 7c** and **d**).

2.3 Toronto, Canada

Toronto is the capital city of the province of Ontario and the home of 2,929,886 residents, 6,346,088 residents within the Greater Toronto Area (GTA)[2] (as of July 2017) [35–37]. A dynamic city, Toronto, is a global city where people of various nationalities and ethnic groups live together. Geographically, Toronto is located on a broad sloping plateau cut by numerous river valleys that cover 641 km^2 and shoreline facing Lake Ontario ,which stretches 43 km (or 138 km when including the bays and islands). As of 2007, GDP of Toronto was 168 US billion dollars and $313.1 US billion dollars in GTA. Toronto is a popular destination to visit; visitors in GTA were recorded as 40 million in 2015 and 93,575 people were counted as nonpermanent residents while 23,065 people aboriginal identity who speak over 60 own languages by report of 2011 statistic Canada [38].

Toronto, with respect to Seoul or Tokyo, experiences much less diversity and scale of natural disasters on average; while Toronto experiences extreme colds or heat alerts [39] during its seasonal extremes; it does not encounter hurricane, earthquake, tsunami, volcano, or flooding on a regular basis. Thus, in comparison, Toronto residents are privileged people in terms of natural disasters.

2.3.1 The emergency plan emphasizes flexibility

The City of Toronto published the Emergency Plan (the Plan, 2017) [40]. In this plan, the city is prepared to deal with any kind of hazards. However, historically, the City of Toronto did not experience natural disasters as much compared with other cities in Canada or similarly scaled cities in the world. Nonetheless, the city of Toronto states that the city is vulnerable to numerous hazards, not only natural disasters such as extreme weather, but also more likely human-caused technological incidents such as those involving hazardous materials, infrastructure disruptions, utility and power failures, or a cyber-attack. It seems that Toronto recognizes that it is more vulnerable to human-caused disasters than to natural disasters from given examples.

The Plan establishes the framework of methodology with necessary documents that the city requires to recover the situation to the normal state by mobilization of resources in case of emergency, and ensure the roles and responsibilities by designing a tool to assist emergency and municipal services. The Plan addresses implementation by flexibly delegating appropriate city agencies roles and authorities in an emergency situation; various supports and services must be provided immediately from other authorities, such as adequate personnel, equipment, and

[2] The Greater Toronto Area (GTA): the central city of Toronto and the four surrounded regional municipalities: Durham, Halton, Peel, and York.

expertise from response agencies. Reviewing and exercising the plan is also required on a regular basis for appropriate implementation. Moreover, most importantly, the Plan emphasizes flexibility to adapt to a broad spectrum of emergencies. Also, the structure of the supporting agencies provides the direction to save lives, protect property and environment, restore essential services and critical infrastructure, and help victims and communities return to normal following an emergency. When Toronto faces an emergency situation, the hazard level will determine its emergency level among four tier emergency level systems as follows:

- Level 0—Normal
- Level 1—Incident
- Level 2—Emergency
- Level 3—Major emergency

When the hazardous situation is determined as a Level 2 Emergency or in the early onset of a Level 3 Major emergency, the operational response team, called the Cluster "B", is called before the Emergency Operations Centre is activated. The Cluster "B" provides strategic management, an operational hub, and a supporting facility for the coordination of emergency response. The Plan clarifies that "once the Emergency Operations Centre is fully activated, the Cluster 'B' operational response team will transition responsibility for operational response to the Emergency Operations Centre to ensure organizational effectiveness and the centralized command is maintained." Also, communication activities such as those outlined in the emergency information and media relations are respected depending on the nature of the emergency.

The Ontario Provincial Emergency Management and Civil Protection Act, Section 2.1 [41] requires that the City of Toronto must develop and implement an emergency management program and adopt it by-law. Training programs and exercises for employees of municipalities, identifying the risk level, and educating the public on risks to public safety and on public preparedness for emergencies are major roles of this act. The Ministry of Community Safety and Correctional Services is committed to ensuring that Ontario's communities are supported and protected by law enforcement and public safety systems that are safe, secure, effective, efficient, and accountable. Their major roles include maintaining the physical and economic security and eliminating hazards to persons or property [42]. The City of Toronto should follow this workflow.

2.3.2 Risks with early warning in Toronto

Environment Canada provides the seasonal weather information to the public with map that illuminates hazard level, warning, watch, statement, or no alert as shown in **Figure 8**. Torontonians can freely access this information to get the climate-related alert. If there is any emergency information with safety issues, both Environment Canada Weather Alert website [43] and Toronto's Alert website provide the latest information to the public, so public can follow the direction to save their life and property. Public health risk and industrial risk are also the City of Toronto's important concerns, and such information is also provided by the city.

The City of Toronto picks up 10 hazards under three categories in 2017, natural hazards, human-caused hazards, and technological hazards; electrical power disruption, explosions/fires, winter weather, pandemic, cyber threats, terrorism,

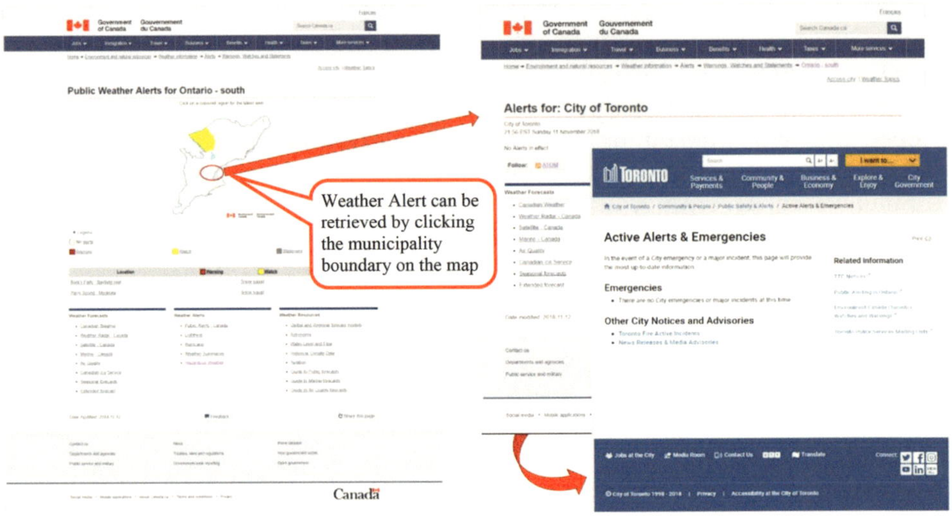

Figure 8.
Public weather alerts for Ontario, which provide real-time watch/awareness, alert, or no alert information to the public directly.

epidemic, flood, fuel/natural gas supply disruption, and extreme heat. Risk assessment is followed by the Ontario's risk assessment matrix (**Table 3**). Risk severity is depicted in different colors (as cited, [44]).

2.3.3 The unique city with diversity

The City of Toronto with diverse races holds 51.5% of visible minority of population, and 51.2% of immigrants were born outside of Canada. Among them, 132,765 people reported that they cannot speak English or French (which are the official language of Canada). This is an important fact to keep in mind, as Toronto is a city that is home to many immigrants and refugees, in addition to many tourists each year. So the City of Toronto must also consider about their vulnerability due to language or cultural barriers. Looking at the data of ethnic demography [45], in 2006, 126 ethnic origins were increased to 278. The different mother tongue (language first learned at birth) users are scattered in Toronto which counted to 89 in 2011, and 85 in 2016; 14 visible minority groups were counted in 2016. Nonvisible minority is indicated as groups from countries in Europe; however, it does not indicate that they use English or French. Some ethnic neighbors are living together in a community that makes Toronto hosts a diverse set of towns such as Koreatown, Little India, Greektown, Corso Italia, Chinatown, and Little Jamaica. The City of Toronto offers to the public its diversity by categories in the 2016 thematic maps where not a visible minority or multiple visible minorities reside as their home, for example

	Probability Rating			
Severity Rating	A. Highly Likely	B. Likely	C. Possible	D. Unlikely
4 – Catastrophic	High	High	Moderate	Low
3 – Critical	High	High	Moderate	Low
2 – Serious	Moderate	Moderate	Moderate	Very Low
1 – Marginal	Low	Low	Very Low	Very Low

Table 3.
Risk assessment matrix of Ontario, which the City of Toronto shall follow as a standard to assess and determine the risk's level.

(**Figure 9**). Understanding the cultural background and language specification is important for both those who give support and evacuation instructions and those who get help in emergency circumstances. Language/cultural risk in emergency situation is another concern for Toronto.

2.3.4 Preparing for a better understanding

The City of Toronto allows the public to access the interactive map, which is called wellbeing index (**Figure 10**) [46]. The city's wellbeing index includes a mapped overlay that shows all community organizations who have agreed to be listed as organizations that have the ability to accept volunteers and provide assistance in the event of an emergency.

2.4 Using mass media for risk communication

The recent trend for urgent information dissemination is using on-line based technology, such as SNS. Government agents are also using SNS such as Facebook (**Figure 11**), Twitter, YouTube, and others [7]. Especially, depending on the people and their culture, specific SNS site is more actively used; for example,

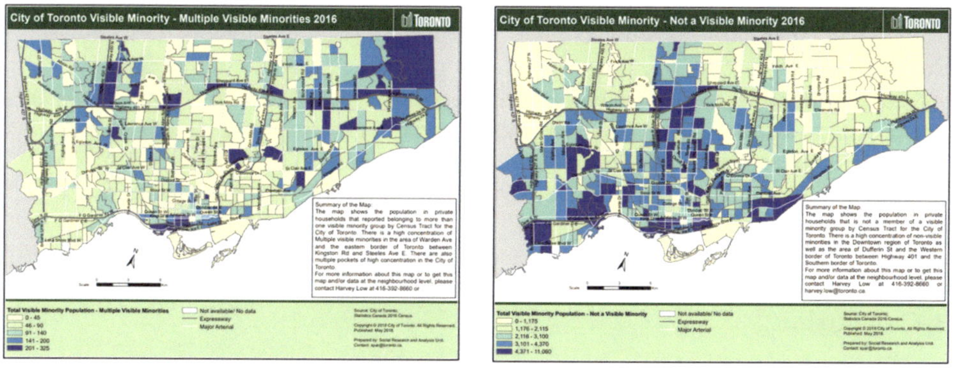

Figure 9.
The 2016 thematic maps where not a visible minority or multiple visible minorities are distributed in Toronto.

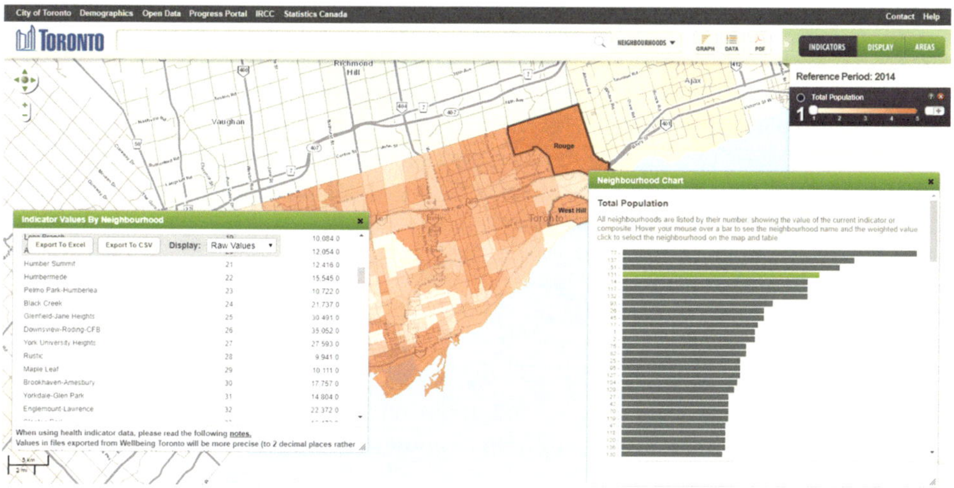

Figure 10.
Wellbeing index (map); information is displayed using query function over spatial map.

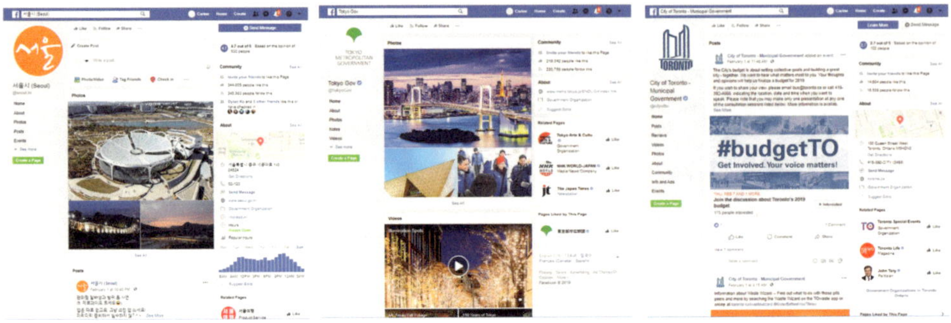

Figure 11.
Each government's official Facebook page: Tokyo (left), Seoul (middle), and Toronto (right). Each city tries to communicate with the public; individuals, agents, volunteers, media, and each other.

Facebook is widely used in the Philippines. As revealed by the 2011 earthquake in Japan and the Philippine Typhoon Haiyan [6, 7], the government, the family and the volunteer groups all used social networking services to communicate information.

Such sophisticated technologies require internet access to use; thus, it would be a fatal weakness for areas with little or no internet, such as remote areas or even urban areas with decommissioned internet. In a highly developed IT city, the disconnection of internet would be a temporary inconvenience, as the public would assume that the service will be recovered quickly because it is considered as one of top priorities. Urban residents, who tend to be more globalized people due to frequent connection to the outside world, use personal IT devices not just to stay connected to the world but also to look for better tools and information to make themselves feel safe and calm.

3. Conclusion

In terms of disaster risk communication, three cities have been developing effective methods and have attempted to provide accurate information to the public. Tokyo, Seoul, and Toronto have different disaster threatening aspects, population characteristics, cultural background, scale, and complexity of the city with physical infrastructures (such as subway system or electricity supply system), so the approach to response to emergencies is differently developed.

In disaster risk management, the disaster cycle has response phase, recovery phase, and preparedness phase. In response phase, there are rescue, supply (materials, equipment, medication, etc.), and support. In recovery phase, there are reopening of the school, repairing of the transportation system, and hard (that is, physical and structural) reconstruction and recovery of infrastructure. In preparedness phase, there are development of manuals, evacuation drills, public education, and so on. Historically, Japanese rely on their government for the entire cycle of a disaster. In the case of Tokyo, despite the many efforts Tokyo makes for the safety of its residents, the information it provides seems to be closer to risk governance than to risk communication. Information that Tokyo provides is mostly preparedness stage information against natural disasters rather than human-related risks, and in general, the government tends to be more prompt with its communication on natural disasters. Therefore, the public most likely depends more critically on broadcasts, media, SNS, and other sources to obtain accurate real-time information on topics about earthquake, tsunami, and typhoon more than risks of terrorism and global political threats.

Won and Kim [47] identified problems by reviewing the disaster and accident scenarios of Seoul and suggested new recommendations for risk communication. By this study, Seoul needs to realize the uncertainties of the disaster incidents and accidents, especially in highly developed complex urban areas. Earthquake and tsunami are not probable natural disasters for Seoul residents, but fine particle dust, such as PM2.5 or yellow sand storm containing heavy metals from China are more realistic risks. Their suggestion emphasizes that initial response is extremely important, and in this early response time stage, every individual should be able to judge the current situation and take their own initiative action. This is a particularly important point in that it can critically affect the life of oneself at an early point in time. On the other hand, the residents in Seoul are dealing with not only high population density, but also complicated transportation system and social services. However, the public is limited in their awareness and knowledge for disaster response in their daily lives—when a crisis occurs, this lack of public awareness/knowledge amplifies the extent of the catastrophe or casualties. Nevertheless, manuals and suggestions are still deployed in governmental agencies and/or organizations, including public and private, and educational sectors.

Considering the hierarchical social work structure in Tokyo and Seoul, top-down control/command, and bottom-up report system, especially, governmental office workers are hired through the employment qualification and hired for a permanent position, but because of the culture of rotating the positions, the work environments hardly accept flexible self-decision-making with one's own responsibility.

Furthermore, there is no manual for residents of Seoul who may not fully understand the government's purpose, intention on disaster drills, or their messages to the public. There may be many well-made drills, policies, and procedures developed by the government, but much of this information do not end up with the public who need these to deal with the emergencies because the channels for the information to reach these end-users are often broken, incomplete, or complex. Thus, these well-developed recommendations and manuals by different government agencies would be "floating in the air", or locked in the storage. This is one of many typical examples of the risk communication being unilateral between decision-makers, field workers, and the victims, which may be a cause of major controversies in bureaucratic administrators.

The unique character of Toronto is the mosaic society with various races and cultures. Still, Canadian government is willing to accept new immigrants from other countries and is open to refugees, as well. The diversity of Toronto society members may characterize the different vulnerability compared to Tokyo and Seoul.

Three cities developed many materials to communicate between governmental agencies and public including manuals and documentary forms, and the clarified relationship between those who command and those who report. According to information on the official websites of the cities, despite the development of IT functions, Toronto seems to have problems communicating with groups with diverse cultural backgrounds and language limitations. Because most information requires a fairly high level of English or French and/or IT knowledge or fluency, it seems that the public experiences difficulty accessing information easily unless particular attention is paid. Moreover, understanding the risks depends on their past experiences, for example, if refugees came from a country in a civil war who will be more aware of the risks of war and terrorism than natural disasters. This can be even more difficult if the amount of information is large. Thus, providing a booklet or pamphlet in different languages is an effective way for public understanding and education depending on the community's ethnic population.

Insurance and financial support allow the victims to return back to their home—their original place—easier and quicker. People returning back to their communities is a very crucial aspect of the society recovering from an event. People return to their communities because of many reasons—memories, assets, land, and other aspects of their lives that are precious to them. This allows the community to rebuild itself, and thus, stand back on its feet and contribute to the society, both economically and culturally. Insurance offered by governmental agencies may potentially give a feeling and sense of safety and hope.

Lessons from 2011 Japan Earthquake and tsunami show that victims were suffering to bring back their original level of quality of life prior to the disaster. Many victims never got their original life back because they did not purchase the earthquake and tsunami combination insurance. The victims could not get the full support from the government up to their expectations. Depending on their original level of quality of life, some victims suffered from the feeling of relative deprivation and decreased financial flexibility. This suggests that mental care of the victims is as important, and having more available options of insurance may reduce or prevent further sufferings of their own people and help them get back on their feet. This highlights an element that is critical in community recovery, resilience, and necessity that were absent in the governance of the three cities.

Depending on the city and their culture, the disaster cycle is often a spiral motion rather than a complete "circle"—as time goes on, as people review and learn from their experiences. As technologies improve, many systems and infrastructures are either repaired or replaced by faster, cleaner, safer, smarter, or better technologies and alternative materials, thus improving the system it replaced with a longer-lasting or more resilient model, leading to a better quality of life for its people and social systems.

Author details

Carine J. Yi[1*] and Tim Park[2]

1 R. Park and Associates Inc., Toronto, Canada

2 PR Immigration and TR Consulting Canada, Toronto, Canada

*Address all correspondence to: cyi.rpa@gmail.com

References

[1] United Nations. The World's Cities in 2016. United Nations; 2017

[2] Renn O. Risk Governance—Coping with Uncertainty in a Complex World, London. Sterling, VA: Earthscan in the UK and USA; 2008. 476 p

[3] Renn O. Risk communication: Insight and requirement for designing succesfull communication programs on health and environmental hazards. In: Heath RL, Dan O'Hair H, editors. Strategies for Overcoming Challenges to Effective Risk Communication. New York: Routledge; 2009

[4] Sjöberg L. Factors in risk perception. Risk Analysis. 2000;**20**(1):1-12

[5] Baram M. The right to know and the duty to disclose hazard information. American Journal of Public Health. 1984;**74**(4):385-390

[6] Yi C, Kuri M, Imamura F, Murao O, Iuchi K, Izumi T, et al. HFA IRIDeS Review Preliminary Report—Hyogo Framework for Action 2005-2015: Building the Resilience of Nations and Communities to Disasters. Japan: International Research Institute of Disaster Science (IRIDeS), Tohoku University; 2013

[7] Yi C, Kuri M. The prospect of online communication in the event of a disaster. Journal of Risk Research. 2016;**19**(7):951-963

[8] Wachinger G, Renn O, Begg C, Kuhlicke C. The risk perception paradox—Implications for governance and communication of natural hazards. Risk Analysis. 2013;**33**(6):1049-1065

[9] Lim S. Acase—[MERS Correspondence 2] Tell the Truth Which We Knew Through Confirmation Process. 2015. Available from: https://acase.co.kr/2015/06/02/mers2/comment-page-1/#comment-11727 [Accessed: July 6, 2015]

[10] Lloyds. Lloyd's City Risk Index 2015-2025. 2015. Available from: https://understandrisk.org/publication/lloyds-city-risk-index-2015-2025/

[11] Seoul Metropolitan Government. City Overview. Available from: http://english.seoul.go.kr/get-to-know-us/seoul-views/meaning-of-seoul/4-population/

[12] The Seoul Institute. Seoul Research Data Service. Available from: http://data.si.re.kr/node/337

[13] Seoul Metropolitan Government. Opened Data Plaza. Available from: http://data.seoul.go.kr/

[14] Timmons H. The World's Largest City Economies. 2018. Available from: https://www.theatlas.com/charts/ryc_Xtn_ [Accessed: November 9, 2018]

[15] Trujillo JL, Parilla J. Redefining Global Cities—The Seven Types of Global Metro Economies. The Brookings Institution, Metropolitan Policy Program; 2016

[16] Statistics Korea, Social Research Results for 2018 (Family, Education, Health, Safety, Environment). 2018. Available from: http://kostat.go.kr/assist/synap/preview/skin/doc.html?fn=synapview371501_1&rs=/assist/synap/preview [Accessed: November 5, 2018]

[17] Coburn A. The Lloyd's City Risk Index 2015-2025—Special Methodology Seminar Lloyd's City Risk Index—Methodology and Usage of City Economy Risk Analysis. 2015. Available from: https://www.jbs.cam.ac.uk/fileadmin/user_upload/research/centres/risk/downloads/150903_

lloydscityriskindexlaunch_coburn_presentation.pdf [Accessed: October 6, 2015]

[18] South Korea News Net. Tycoon Wanted in Fatal South Korean Boat Capsize Found Dead. 2014. Available from: https://www.southkoreanews.net/news/224021035/tycoon-wanted-in-fatal-south-korean-boat-capsize-found-dead [Accessed: November 9, 2018]

[19] Park M, Hancocks P. Sewol Ferry Disaster: One Year on, Grieving Families Demand Answers; CNN. 2015. Available from: https://www.cnn.com/2015/04/15/asia/sewol-ferry-korea-anniversary/index.html [Accessed: November 9, 2018]

[20] The Editorial Board of the New York Times. Duty and Shame as the Ship Sank; The New York Times. 2014. Available from: https://www.nytimes.com/2014/04/23/opinion/duty-and-shame-as-the-ship-sank.html

[21] Mullen J, Kim S. South Korea Cracking Down on Operator in Sewol Ferry Disaster; CEO Arrested; CNN. 2014. Available from: https://www.cnn.com/2014/05/08/world/asia/south-korea-ship-sinking/index.html

[22] City of Seoul, Seoul Crisis Management Communication Guidelines. 2014. Available from: https://opengov.seoul.go.kr/public/5349666

[23] Seoul Metropolitan Facilities Management Corporation. Safety Department, Strengthening the Reporting System for Disasters; 2016

[24] National Emergency Management Agency of Korea. Disaster Preparedness Training Manual; 2013

[25] Choe S. South Korea's Leader and Media Face Scrutiny Over Ferry Disaster; The New York Times. 2014. Available from: https://www.nytimes.com/2014/05/10/world/asia/south-koreas-leader-and-media-face-scrutiny-over-ferry-disaster.html

[26] Seoul Safety Nuri Portal Site. Available from: http://safecity.seoul.go.kr:8070/scmyn_cf/map/cfMap.do?type=hewaRefuge#

[27] Water Journal (Online). Future Sewer Policy Direction of Seoul City [Issue & Trend]. 2017. Available from: http://www.waterjournal.co.kr/news/articleView.html?idxno=38539 [Accessed: September 28, 2017]

[28] Metropolitan Government Tokyo. Statistics Division, Bureau of General Affairs, Population of Tokyo. Available from: http://www.metro.tokyo.jp/english/about/history/history03.html

[29] Nikkei. The Budget of Tokyo is 13 trillion yen. Equivalent to Sweden. 2016. Available from: https://www.nikkei.com/article/DGXKZO05133170S6A720C1CC0000/ [Accessed: November 7, 2018]

[30] Lloyd's. Emerging Risk Report 2017. Understanding risk—Reimagining History Counterfactual Risk Analysis. Available from: https://www.lloyds.com/~/media/files/news-and-insight/risk-insight/2017/reimagining-history.pdf

[31] Hiroi U, Sekiya N, Nakajima R, Waragai S, Hanahara H. Questionnaire survey concerning stranded commuters in metropolitan area. Journal of Social Safety Science. 2011;**15**:343-353

[32] Hiroi U, Oomori T, Shinkai H. Evacuation simulation in metropolitan area and risk maps of heavy traffic and crowd in catastrophic disaster. Journal of Japan Association for Earthquake Engineering. 2016;**16**(5):5_111-5_126. DOI: 10.5610/jaee.16.5_111

[33] Wendling C, Radisch J, Jacobzone S. The Use of Social Media in Risk and

Crisis Communication, OECD Working Papers on Public Governance; No. 24. OECD Publishing; 2015. DOI: 10.1787/5k3v01fskp9s-en

[34] Tokyo Disaster Prevention Website. Available from: http://www.bousai.metro.tokyo.jp/foreign/english/kitaku_portal/2001064/2001072.html

[35] Statistics Canada. Census of Population. 2016. Available from: https://www150.statcan.gc.ca/n1/pub/11-627-m/11-627-m2017025-eng.htm

[36] City of Toronto. Available from: www.toronto.ca/

[37] Canada Population. Available from: https://www.canadapopulation.org/toronto-population/

[38] City of Toronto. Toronto at a Glance. Available from: https://www.toronto.ca/city-government/data-research-maps/toronto-at-a-glance/

[39] City of Toronto. Public Safety & Alert. Available from: https://www.toronto.ca/community-people/public-safety-alerts/emergency-preparedness/before-an-emergency/

[40] The City of Toronto. The Emergency Plan; 2017

[41] Ontario Emergency Management and Civil Protection Act. R.S.O. Chapter E.9 (From December 15, 2009 to the e-Laws currency date). 1990. Available from: https://www.ontario.ca/laws/statute/90e09

[42] The Ministry of Community Safety and Correctional Services, Ontario. Guide to Incident Management System (IMS) Implementation. Available from: https://www.emergencymanagementontario.ca/english/home.html

[43] Public Weather Alerts for Ontario. Available from: https://weather.gc.ca/warnings/index_e.html?prov=son [Accessed: November 11, 2018]

[44] Public Services Health & Safety Association and Toronto Central Local Health Integration Network (LHIN). A Guide to an Emergency Management Plan; 2014

[45] City of Toronto. Toronto Social Atlas. Available from: https://www.toronto.ca/city-government/data-research-maps/neighbourhoods-communities/toronto-social-atlas/

[46] City of Toronto. Community Partners on the Wellbeing Index. Available from: https://www.toronto.ca/community-people/public-safety-alerts/emergency-preparedness/during-an-emergency/community-partners-on-the-wellbeing-index/

[47] Won J, Kim S. A Study of Developing Disasters and Incidents Scenarios Based on Emergency Responses by The Seoul Institute; 2017

Section 2

Risk, Public/Private and Financial Services Contex

Chapter 4

Bank Risk Management: A Regulatory Perspective

Nguyen Thi Thieu Quang and Christopher Gan

Abstract

The globalization of financial markets, information technology development, and increasing competition have largely affected bank business and its risk management. Together with these forces, regulatory factors play a significant role. This chapter approaches bank risk management under the regulators' perspective with an emphasis on the risk-based capital regulation. Specifically, how bank risk is regulated under the risk-based capital regulation and whether the regulation shapes bank risk are discussed in detail. In such a way, the chapter provides better understanding of the risk-based capital regulation and bank risk-taking behaviors.

Keywords: Basel Accords, capital regulation, bank risk, risk management, credit risk

1. Introduction

Risk management is important for a bank to ensure its profitability and soundness. It is also a concern of regulators to maintain the safety and soundness of the financial system. Over the past decades, banking business has developed with the introduction of advanced trading technologies and sophisticated financial products. While these advancements enhance bank's intermediation role, promote profitability, and better diversify bank risk, they raise significant challenges to bank risk management. The risk management of banks has been considered to be weak compared to the rapid changes in the financial markets [1]. In the light of the recent global financial crisis, bank risk management has become the major concern of banking regulators and policy makers.

Basel Committee for Banking Supervision (hereafter, the Basel Committee), which was established in 1974 by the central bank governors of G-10 countries[1], acts as the primary global standard setter for banking prudential regulation. The Committee sets international standards and guidelines for national regulators to assess and supervise their banking system. Its landmark publication—the Basel Accord—largely affects the way banks manage their capital and risk as well as the way they are monitored and supervised by the regulators.

This chapter approaches bank risk management under the regulators' perspective with an emphasis on the risk-based capital regulation. Section 2 provides a brief overview about bank risk and risk management. Since the regulation practice in most countries is pretty much based on the guidelines of the Basel Committee, this section primarily follows the Committee's documents. Section 3 introduces

[1] The Group of Ten (G-10) is made up of 11 industrial countries, including Belgium, Canada, France, Germany, Italy, Japan, the Netherlands, Sweden, Switzerland, the UK, and the USA.

the Basel Accords (i.e., Basel I, II, and III), which set the international standards for bank capital regulation, and clearly states how and what types of bank risk are covered by the Accords. Although the Accords, and the capital regulation in general, are expected to discipline the risk-taking behavior of banks, such effect is still under debate. Sections 4 and 5, respectively, present theoretical arguments and global evidences about whether the capital regulation reduces bank risk-taking. For the purpose of the chapter, only risk-based capital regulation is concerned. Section 6 provides a summary of the chapter.

2. Overview of risk and risk management in banking

Bank risk is usually referred as the potential loss to a bank due to the occurrence of particular events. Key risks in banking include credit risk, interest rate risk, market risk, liquidity risk, and operational risk.

Credit risk is "the potential of a bank borrower or counterparty that will fail to meet its obligations in accordance with agreed terms" [2]. Exposure to credit risk is the largest and major source of problems in most banks. Credit risk does not only derive from loans but also from other activities on both banking book and trading book, as well as on- and off-balance sheets. Therefore, credit risk also comprises counterparty risk (the risk that a party in a financial transaction will default).

Interest rate risk (in the banking book) is related to the adverse movements in interest rates of bank assets, liabilities, and/or off-balance sheet items. A change in the interest rate affects a bank's expected interest incomes and expenses and thus affects its future marginal profits. The Basel Committee [3] identifies three main types of interest rate risk including gap risk, basic risk, and option risk. Gap risk arises from the term structure of the interest rates. Basic risk is related to the relative changes in interest rates for financial instruments with similar tenors but priced using different interest rate indices. Option risk, on the other hand, arises from option derivatives or from options embedded in a bank's assets, liabilities, and/or off-balance sheet items. This third type is indirect interest rate risk because it depends on the decisions of the bank and/or its customers.

Market risk is "the risk of losses in on and off-balance-sheet positions arising from movements in market prices" [4]. This definition covers both interest rate risk related to instruments and equity marked to market, foreign exchange risk of positions in foreign currencies, and the price risk of commodities that can be traded on a secondary market.

Liquidity risk refers to the bank's inability to fund the increase in assets and meet obligations at a reasonable cost. Two types of liquidity risk include funding liquidity risk and market liquidity risk. Funding liquidity risk is the risk that the bank cannot meet efficiently current and future cash flow and collateral needs without affecting its daily operations or the financial condition. Market liquidity risk is the risk that the bank is unable to easily offset or eliminate a position at the market price due to inadequate market depth or market disruption [5]. Liquidity risk is inherent in banking since there is usually a maturity mismatch related to bank's transformation of short-term liabilities into longer-term assets.

Operational risk is "the risk of loss resulting from inadequate or failed internal processes, people and systems or from external events" [6]. Operational risk can arise from any banking products, activities, processes, and systems.

Risk management is important for banks to ensure their profitability and soundness. It is the process established by bank managers to ensure that all risks

Types	Documents	First issued	Last revised
Standards	Core principles for effective banking supervision	Sep 1997	Sep 2012
Guidelines	Corporate governance principles for banks	Sep 1999	Jul 2015
Guidelines	Principles for sound stress testing practices and supervision	May 2009	None
Standards	Basel III: A global regulatory framework for more resilient banks and banking systems	Dec 2010	Jun 2011
Standards	Basel III: The liquidity coverage ratio and liquidity risk monitoring tools	Dec 2010	Jan 2013
Standards	Basel III: The net stable funding ratio	Dec 2010	Oct 2014
Standards	Basel III: Finalizing post-crisis reforms	Dec 2017	None
Standards	Basel II: International Convergence of Capital Measurement and Capital Standards: A Revised Framework—Comprehensive Version	Jun 2004	Jun 2006
Guidelines	Principles for the Management of Credit Risk	Sep 2000	None
Guidelines	Best Practices for Credit Risk Disclosure	Sep 2000	None
Guidelines	Principles for Sound Liquidity Risk Management and Supervision	Feb 2000	Sep 2008
Guidelines	Principles for the Sound Management of Operational Risk	Feb 2003	Jun 2011
Standards	Interest rate risk in the banking book	Sep 1997	Apr 2016

Source: Authors' compilation from Basel Committee's publications (available at https://www.bis.org/bcbs/publications.htm?m=5%7C28%7C427).

Table 1.
Basel Committee documents on bank risk management.

associated with the bank's activities are identified, measured, limited, controlled, mitigated, and reported on a timely and comprehensive basis [7]. A sound risk management system is necessary to support the regulators in assessing bank's soundness and to reinforce the market participants' confidence in the banking system. The Basel Committee [1] suggests that a sound risk management system should have: (i) active board and senior management oversight; (ii) appropriate policies, procedures, and limits; (iii) comprehensive and timely identification, measurement, mitigation, controlling, monitoring, and reporting of risks; (iv) appropriate management information systems at the business and firm-wide level; and (iv) comprehensive internal controls.

The Basel Committee sets out a number of standards and guidelines for bank risk management. **Table 1** summarizes main documents related to regulation of bank risk management by the Basel Committee.

3. Regulation of bank risk under the Basel Accords

Bank risk management and capital management are inseparately related to each other. As specified under Principle 15 in Core Principles for Effective Banking Supervision by the Basel Committee [8], the bank's risk management process should be able to assess the adequacy of their capital in relation to their risk profile as well as market and macroeconomic conditions. In a similar way, Principle 16 requires that the capital adequacy requirements should reflect all risks taken and presented by a bank given the markets and macroeconomic conditions that it operates. In such manner, the Basel Accords specify capital requirements to adequately cover bank risks. The following provides details of how and what types of risks are regulated under Basel Accords.

3.1 Basel I

The risk-based capital adequacy standards were introduced by the Basel Committee in 1988 and commonly known as Basel I Accord. The Accord not only sets standards for capital adequacy but also affects the bank's risk management. Initially, only credit risk is addressed since it is the major risk of banks. Under Basel I, banks are required to hold an adequate amount of capital to cover their credit risks related to different categories of on- and off-balance sheet assets. For on-balance sheet assets, the Accord assigns a risk weight ranging from 0, 10, 20, and 50 to 100% on different asset categories. For off-balance sheet exposures, credit conversion factors are applied to different types of off-balance-sheet instruments or transactions based on their estimated size, the likely occurrence of the credit exposure, and the relative degree of credit risk. These credit conversion factors are then multiplied by the weights applicable to the category of the counterparty for an on-balance-sheet transaction to determine the amount of risk-weighted assets [9].

Efforts to address other risks other than credit risk have been made continuously. As a result, Basel I was refined in January 1996 to comprise requirements for market risk which arose from exposures to foreign exchange, traded debt securities, equities, commodities, and options. In measuring market risk, banks are allowed to choose between standardized approach and internal models approach, subject to the approval of the national authorities. By incorporating market risks into the risk framework, Basel Committee clarifies that credit risk requirements exclude debt and equity securities in the trading book and all positions in commodities but include the credit counterparty risk on all over-the-counter derivatives whether in the trading or the banking books [10].

3.2 Basel II

The framework was revised in June 2004, commonly referred as Basel II, with the introduction of operational risk in addition to the existing credit risk and market risk. Under the Basel framework, operational risk includes legal risk but excludes strategic and reputational risk. Basel II provides three available methods to calculate capital charges for operational risk, including basic indicator approach, standardized approach, and advanced measurement approach. The use of standardized approach and advanced measurement approach is subject to supervisory approval [11].

The revised framework also provides more flexibility in calculating capital requirements for credit risk. Accordingly, banks are allowed to choose between standardized approach and internal rating-based approach, depending on which approach is more appropriate for their operations and financial market infrastructure. The credit risk measurement is also more risk sensitive than in Basel I. Specifically, standardized approach is supported by external credit assessments, and the risk weights have a wider range from 0 up to 350% (for the case of securitization tranches that are rated between BB+ and BB−). There is also consideration for past due loans more than 90 days, treatments of credit risk mitigation, and securitization exposures [11].

Despite attempting to capture most risks in the minimum capital requirements, Basel Committee acknowledges that it is impossible to cover all risks. For such reason, they introduce Pillar 2 for Supervisory Review and Pillar 3 for Market Discipline to complement the minimum capital requirement, which is regarded as Pillar 1.

The Pillar 2 aims to treat risks that are not fully included or considered in Pillar 1 such as credit concentration risk, interest rate risk in the banking book, business and strategic risk, and factors external to the bank such as business cycle effects. In addition to ensure that banks have adequate capital to support their risks, Pillar 2 encourages banks to develop and use better risk management techniques. Therefore, banks are required to have an overall capital adequacy assessment process according to their risk profile and a capital maintaining strategy. This involves the board and senior management oversight, a sound capital assessment, a comprehensive assessment of risks, monitoring and reporting, and an internal control review. Bank's self-assessment will be then reviewed and evaluated by supervisors [11].

The Pillar 3 complements the other two pillars by requiring banks to disclose information about their scope of application, capital, risk exposures, risk assessment processes, and the capital adequacy. The disclosure is expected to inform the market about the bank's risk exposures and provide a consistent and understandable framework that enhances the comparability among banks [11].

The 2004 Basel framework focuses primarily on the banking book. More attention to trading book as well as exposures to the *double default* is expressed in a consensus document released in July 2005 [12]. Accordingly, the Basel Committee provides detailed treatment of counterparty credit risk and cross-product netting, treatment of double-default effects, short-term maturity adjustments in the internal rating-based approach, improvements to the current trading book regime, and specific capital treatment for failed transactions and transactions that are not settled through a delivery-versus-payment framework. These changes ensure the risk sensitivity of capital requirement for credit risk and market risk in Pillar I, as well as enhance the requirements for bank's internal capital adequacy assessment and market disclosure, particularly for the counterparty credit risk, concentration credit risk, and trading book. The document was then incorporated into the existing framework in June 2006 [4].

3.3 Basel III

The massive failure of the banking system during the global financial crisis 2007–2009 forced the revision of Basel II. Consequently, Basel III was issued in mid-December 2010 and consequently revised in June 2011. The new framework enhances the risk coverage in three Pillars, particularly credit risk and market risk by raising capital requirements for the trading book and complex securitization exposures, introducing a stressed value-at-risk (VaR) capital requirement based on a continuous 12-month period of significant financial stress, requiring higher capital charges for so-called resecuritizations in both the banking and the trading books and strengthening the capital requirements for counterparty credit exposures arising from banks' derivatives, repo, and securities financing activities [13].

Basel III also introduces liquidity requirements to address bank's liquidity risk, which was considered as not properly managed during the early phase of the financial crisis in 2007. Two minimum standards for funding, namely, liquidity coverage ratio (LCR) and the net stable funding ratio (NSFR), are proposed. These two standards are designed to promote both short-term and long-term resilience of a bank's liquidity risk profile. Specifically, the LCR ensures that the bank has sufficient high-quality liquid resources to survive an acute stress scenario lasting for 1 month. And the NSFR creates additional incentives for a bank to fund its activities with more stable sources of funding on an ongoing structural basis (1 year). In supporting supervisors to monitor banks' liquidity risk profiles, Basel Committee develops a set of common metrics to be considered as the minimum types of information for supervisors. These include contractual maturity mismatch, concentration of funding, available unencumbered assets, LCR by currency, and market-related monitoring tools [13]. The regulation on liquidity risk was revised in January 2013 for the liquidity coverage ratio and liquidity risk monitoring tools and in October 2014 for the net stable funding ratio [14, 15].

In December 2017, the Basel Committee completed its Basel III post-crisis reforms. Accordingly, the revised framework enhances the robustness and risk sensitivity of the standardized approaches for credit risk and reduces the reliance of its external credit ratings. The use of internally modeled approaches is more constrained. For operational risk, the framework replaces the advanced measurement approaches and the existing three standardized approaches with a single risk-sensitive standardized approach. The new standardized approach is determined based on bank's income and bank's historical losses [16].

4. The effect of risk-based capital requirements on bank risk behavior

The risk-based capital requirements aim at creating a discipline for bank risk behavior, through which they help ensure the global financial stability. However, there are controversies whether they shape bank risk. Since the introduction of Basel Accord in 1988, continuing efforts have been concentrated on showing how these risk-based capital requirements affect bank risk. Studies fall into four theories, which focus on different factors to explain bank risk behavior. These factors include moral hazard, franchise value, capital buffer, and agency problem.

4.1 Moral hazard theory

Given the limited capital and the existence of information asymmetry, banks are induced to take excessive risk. This is often regarded as the moral hazard problem, whereby one party to a transaction engages in activities detrimental to the other party [17]. The problem is exacerbated with the existence of the deposit insurance. In order to improve the depositors' confidence in the banking system and stabilize the financial system, the government usually provides banks with the deposit insurance (either implicitly or explicitly). Under the protection of deposit insurance, the depositors incur no risk of depositing into the bank. Thus, they lose incentives to monitor bank operations. Consequently, banks have incentives to take greater risks. Therefore, researchers examining the effect of risk-based capital requirements on bank risk-taking in a context of the moral hazard focus on the limited liability of banks and the existence of the deposit insurance. However, using different models and assumptions, these studies arrive at different conclusions. Negative effect of capital requirements on bank risk-taking is suggested by Kim and Santomero, Flannery, Furlong and Keeley, Gjerde and Semmen, and Zweifel, Pfaff, and Kühn [18–22]. However, Gennotte and Pyle and Rochet suggest a positive effect [23, 24]. Mixed result is proposed by Calem and Rob, Blum, and Silva [25–27].

Kim and Santomero [18] adopt a single-period mean-variance model, which assumes that banks are single-period risk-averse expected utility maximizers.

The bank's objective function is to maximize the utility, which depends on the mean and standard deviation of return on equity. The capital requirements will set a bound on the bank's expected return on equity and alter its choices of risk and return. Consequently, bank hesitates to extend credit to assets with high-risk weights and shuffles the asset portfolio toward more safe assets and less risky assets.

Flannery [19] explains the effect of risk-based capital requirements on bank risks by unifying the above utility function with an option price function. The option price function regards deposit insurance as a put option, which the bank can maximize by selecting the riskiest available asset portfolio. The model assumes that bank asset decision is two periods and bank chooses a loan portfolio that maximizes its expected return on equity. This return is affected by the value of the deposit insurance option, which depends on the permissible leverage. The capital requirements penalize higher risk with more capital, which imply a negative relationship between risk and leverage. In such way, the capital requirements lower banks' preferred loan risk while they pursue high portfolio risk to maximize the deposit insurance's put option value.[2]

Furlong and Keeley [20] employ both state-preference model and option model. In the state-preference model, the authors consider an insured bank that aims to maximize its current value of equity through two possible future states. With a given initial capital, the bank that seeks to maximize the current value of its equity will try to maximize the value of the deposit insurance option. This is equivalent to the option model, in which the bank's objective is to maximize the value of the deposit insurance. This can be done by maintaining the highest degree of leverage (i.e., the lower capital ratio) allowed by the regulation and increasing asset portfolio risk. The gain from increasing asset risk depends on the asset size rather than the leverage. However, the change in leverage directly affects the asset size. Therefore, the marginal gain from increasing asset risk is positively correlated to the leverage change. This means that lower leverage caused by the capital requirements will reduce the bank's marginal gain from increase in asset risk and, thus, reduce bank's incentive to increase asset risk. In addition, for a given level of leverage, the insurance subsidy's value is positively correlated with asset volumes. This suggests that bank prefers to increase capital rather than shrinking assets and retiring deposits to reduce the leverage. In such way, they can maximize their asset volume and, thus, the value of the insurance subsidy.

Using the same two-period state-preference model, Gjerde and Semmen [21] extend Furlong and Keeley's study [20] by examining both risk-based capital requirements and leverage restriction. They show that under the leverage restriction, bank managers maximize both the leverage up to the regulator's restriction and the asset risk to exploit the most benefit of the deposit insurance. However, the total value of the bank is a negative function of the risk-based equity ratio. The higher risk-based equity ratio reduces the possible investment in riskiest asset and, thus, lowers the value of deposit insurance. Therefore, with a constrained amount of equity, banks will minimize the leverage to be able to invest in high-risk assets. When the amount of equity is not constrained, bank managers will increase the equity until all funds are invested into the riskiest assets. The authors add that either minimum leverage ratio or risk-based capital ratio is efficient in regulating bank risk if the risk weights are optimal. However, if there is suboptimal risk weights (which is very likely in practice), a combination of both requirements is necessary to control bank portfolio risk.

[2] This is because bank portfolio choice not only depends on the regulation but also other factors such as monopoly power, liquidity, scale or scope of economies, risk aversion, or special information availability.

Nevertheless, the reducing-risk effect of capital requirements is not always supported. Gennotte and Pyle [23] depart significantly from previous research by relaxing the assumption about bank assets as zero net present-value projects and show that under certain circumstances, increased risk-based capital requirements lead to higher asset risk. Arguing that evaluating and monitoring loans are costly, the authors assume that bank assets are nonzero net present-value investments. The bank determines its optimal asset portfolio based on the tradeoff between the portfolio's net present value and subsidy value. Bank's response to the increase in capital requirements depends on the sign of the net present value of the asset portfolio, which is a function of the asset risk and the level of investment. Although the market value of bank equity decreases with an increase in asset risk under tightened capital requirements, if the net present value of the asset portfolio is sufficiently negative and large, bank can offset this decrease by increasing the risk and reducing the level of investment simultaneously. Thus, bank asset portfolio comprises both relatively safe assets and risky assets with higher fraction of riskier assets.

Taking into consideration the limited liability of banks, Rochet [24] claims that the constraint in capital ratio can induce banks to choose very "extreme" asset portfolios with specialization on some assets. This negative effect eventually dominates the risk aversion characteristic, and even correct risk weights cannot prevent banks from inefficiently asset allocations.

Arguing that static framework, as in aforementioned studies, might neglect the intertemporal consequences of risk-taking behaviors of banks and preclude cross-sectional predictions of banks with different capital positions, Calem and Rob [25] develop a dynamic model which allows for the variation of bank's capital position over time and across banks to predict bank risk-taking behavior under the capital regulation. In this model, banks are considered to operate in a multi-period setting and aim to maximize the discounted value of their profits. In each period, based on a specific capital position, the bank will determine its portfolio by choosing between its safe and risky assets. From this portfolio choice, the bank's returns are realized. This realization of returns together with the preexisting capital position would then determine the next period's capital position. The process is carried out in such a way that the bank faces the same portfolio choice with different capital positions in each period and equivalently for different banks with different capital positions. Calibrating the model using empirical data on the US banks in 1984–1993, the authors show that under increasingly stringent capital requirements, the level of bank's risk-taking depends on the bank's current capital position with a roughly U-shaped relationship, and bank risk-taking is restrained only if the risk-based capital standard is stringent enough.

In terms of the dynamic of capital regulation, Blum [26] analyzes the effect of capital regulation on bank risk-taking in a multi-period framework. The author shows that if the bank faces capital requirement only in the first period, it would decrease the risk due to the increase in equity. However, tightening capital ratio in the second period generates two possible effects. First, it lowers the expected profit of the banks. Bank managers are induced to take more risk with the perception that they have less to lose in the case of bankruptcy. Second, the binding regulation increases the marginal return on risk and reinforces the first effect. Consequently, the overall risk of regulated banks goes up. Silva [27] completes Blum's model by providing the computed values of the threshold requirements for which the risk chosen by the bank converges to the zero bankruptcy cost and social optimum. These values, in turn, depend critically on the initial equity of the bank. The author confirms that constant capital requirements could efficiently reduce bank's risk-taking and thus achieve the zero bankruptcy cost as well as socially efficient level of risk. However, this effect requires a very high level of capital requirement, which might not be practicable.

Recently, Zweifel et al. [22] consider bank's objective as maximizing the risk-adjusted return on capital (RAROC). The model assumes that bank's optimal capital level is determined through a three-stage process. In the first stage, the bank faces exogenous shocks on expected return and volatility, which affects its solvency-level adjustment in the second stage in a way of maximizing the RAROC. In the third stage, the bank rebalances its assets in response to changes in the solvency level by choosing new values for expected returns and risk. An internal efficiency frontier is formed with its slope depending on the capital regulation such as Basel I and Basel II. Against the expectation that these regulations can reduce the slope of the efficient frontier (i.e., banks choose lower expected return and volatility), the study shows that both Basel I and Basel II may lead banks to choose higher-risk positions than it would otherwise. The risk is likely to increase even in the case of Basel III regulation.

4.2 Franchise value theory

The capital requirements are expected to reduce bank's excessive risk-taking. However, analysis under the moral hazard theory does not always support this argument. A substantial part of researchers has relied on the franchise value to explain the effect of risk-based capital requirements on bank risk-taking. Franchise value is the accumulated present value of a bank's expected future profits if it operates continuously and represents an opportunity cost if the bank goes bankrupt [28]. The higher the franchise value, the more the bank stands to lose by becoming insolvent. In contrast, with no franchise value, the bank has nothing to protect and no worry about bankruptcy. Two main sources of bank's franchise value are *market-related* and *bank-related* factors [29]. *Market-related* factors such as competition environment, legislation restrictions, and technology innovation create the differences in franchise value in banks across geographic or product markets, while *bank-related* factors can originate from efficiency variations in bank operations, relationship management, or branch networks. Researchers under franchise value theory usually take into account the competition of the banking environment to explain the effect of capital requirements on bank risk.

With the notion that competition contributes to the erosion of bank's franchise value and reduces its motivation to take less risk, Hellmann [30] investigates the effect of capital standards in the environment of competition. They argue that capital requirements reduce bank's moral hazard by putting their equity at risk, which they regard as the capital-at-risk effect. However, they can have adverse effect by harming bank's franchise value due to lower per-period future profits and thus induce them to take more risk. The latter consequence is known as franchise-value effect. Liberalization will intensify the competition among banks and encourage them to offer inefficient deposit rates to steal shares from their competitors. In such event, bank can only increase their franchise value by gambling. Therefore, capital requirements in a competitive deposit environment cannot make banks pursue a prudent investment strategy. The authors then suggest that a combination of deposit rate ceiling and capital requirement will help address the problem.

This effect is reexamined by Repullo [31] but in an explicit model of imperfect competition. The author discovers that the Hellmann's conclusion [30] is only true in the case of a very competitive deposit market where intermediaries can earn low return margins. In markets where banks can earn an intermediate margins, they can invest in both risky and safe assets. In the extreme case of monopolistic markets, only prudent investments exist. Moreover, if the cost of capital due to the increase in capital requirements exceeds the returns of the safe asset, capital requirements are always effective in preventing banks from taking excessive risks. This is because banks can fully transfer all the cost of higher capital requirements to the depositors. This makes the equilibrium expected margins unchanged and so does the franchise values. In such case, the increased capital reduces the equilibrium deposit rate in a way that the bank's franchise value does not vary and, thus, reduce bank's incentive to take very risky assets.

Following Repullo's approach [31], Zhang et al. [32] show that bank franchise value decreases with an increase in the capital ratio. With the enforcement of capital requirements, the bank holds a capital ratio as near the minimum capital requirement as possible to maximize its franchise value. In maximizing the franchise value, bank invests less in risky assets with an increase in the capital ratio. Consequently, capital requirements are effective in changing bank risk preference and reducing bank's incentive to take risk.

Behr [33] also examines how stricter capital requirement affects bank's risk-taking in different market structures. The results bear a slight resemblance to Repullo [31] except for the case when the bank operates in a moderate competitive environment. The difference may be in the assumption of the objective functions of the bank. Behr [33] shows that in low concentrated markets, banks have low franchise values. The bank's objective in such an environment is to maximize the short-term profits and, thus, have great incentives to take risks to increase the franchise values. Therefore, capital requirements will play the discipline role to reduce bank's risk-taking. On the other hand, in highly concentrated markets, the banks do not have to compete severely with each other, and their franchise values are higher. The bank's objective now is not only to maximize the short-term profits but also the expected future profits, which are the franchise values. Bank, therefore, will be less induced to engage in high-risk assets as they would threaten its high franchise values. The role of capital regulation in this case becomes less clear.

4.3 Capital buffer theory

Extending from the franchise value literature, emerging studies have focused on the dynamic of bank's franchise value, which forms a new theory of bank behavior under capital regulation—capital buffer theory. Accordingly, there are costs in changing the level of capital and falling below the required capital level. These costs can be implicit or explicit. Implicit costs can arise from the regulatory intervention to limit the likelihood of a deposit insurance, whereas explicit costs refer to the regulators' restrictions or penalties due to noncompliance with the minimum capital requirement or even liquidation [34]. In order to avoid these costs, banks have incentives to hold a buffer above the minimum capital requirement.

Taking into account this incentive effect of capital regulation, Milne and Whalley [35] show that bank's attitude toward risk depends on its capital buffer and that in the long run, the capital regulation has no impact on bank risk behavior. However, this is true only when deposit is not remunerated. When allowing for deposit repayment, an increase in capital requirements increases bank's franchise value[3] and affects the desired capital buffer. The risk-taking incentive

[3] This is because capital is viewed as an endogenous response to capital regulation. This means that an increase in capital constraint leads to bank recapitalization either through reducing dividend or issuing new equity. Although the shareholder value declines, the present value of all future expected payments to shareholders is higher. Therefore, higher capital requirements are associated with higher franchise value.

is thus reduced. But if the bank has adequate earnings, capital requirements exert little impact on bank risk-taking. In the short run, the bank risk behavior is similar to the prediction of franchise value theory. Specifically, fully capitalized banks, which have successfully built up the desired level of capital, are insured against the cost of recapitalization and liquidation, so they aim to minimize the cash flow uncertainty and be risk-averse. However, if those banks suffer from severe deterioration of cash flow but not so much to destroy the value of shareholder, they will take greater risk to avoid costly equity issuance. Whereas, banks with less than minimum required capital (normally when regulatory audit is random) are under the threat of regulation intervention and thus, become risk lovers to maximize the cash-flow uncertainty.

Milne [36] examines this incentive effect of capital requirements on bank's portfolio choice. The author shows that in the short run, banks struggling to meet the regulatory capital requirements will reduce the holding of highly risky assets, while well-capitalized banks face little pressure from regulatory in allocating their portfolio. An exceptional case is failing banks which consider the capital regulation breaches as unavoidable and choose their portfolio without regarding the effect of the regulation. However, to the extent that the value of bank assets can be realized (e.g., through loan trading or securitization), the risk-based capital requirements have no impact on bank's portfolio choice. In the medium term, banks raise capital until the marginal expected cost of breaching the regulatory requirement equals the marginal financing cost of equity and debt. The effect of capital requirements on bank's asset portfolio, thus, depends on the marginal costs of debt and equity finance.

4.4 Agency problem theory

While higher capital level can help reduce the conflicts between the bank's shareholders and debt holders, in this case the depositors, it may reinforce the conflicts between the shareholders and the managers [37]. In banking, the shareholders normally delegate the operations of the bank to the managers. Both parties aim to maximize their benefits, and it would be difficult for the shareholders to ensure that the managers are acting in the best interests of the shareholders. The agency problem refers to the divergence in the interest of managers and shareholders when managers indulged to maximize their own utility rather than the bank value [38].

Besanko and Kanatas [39] argue that the underpriced deposit insurance adds to the bank's surplus from lending. This motivates bank's managers to manage the loans efficiently in order to realize the surplus (i.e., when loans are repaid). However, increasing capital requirements lead to higher cost of fund which reduces the managers' surplus. Together with the issuance of new equity to satisfy the higher capital standards, the insiders' ownership is diluted sufficiently to reduce their incentives to make effort in monitoring the loans. This negative effect of enforced capital requirements can be greater than the benefit of asset substitution by requiring the shareholders to have more capital. The net effect is a rise in the overall riskiness of bank assets.

Taking the effect of general equilibrium into concern, Gale [40] also suggests that increasing capital requirements may have adverse effect on bank risk-taking. Given the nature that the manager's private benefit can be damaged if the bank goes bankruptcy, the managers have incentives to be risk-averse. They will aim to maximize the probability of the success state of the investment subject to the bank's capital constraints. Hence, they will choose the asset with the lowest return but larger than the case without capital constraint. Given the high-risk, high return assumption, the bank risk will increase.

Incorporating the difference in manager's incentives with those of the shareholders and deposit insurers in a model with four distinct characteristics on the risk-return asset profiles, Jeitschko and Jeung [41] show that under capital regulation, the bank risk varies with the relative forces of these agents. If the shareholder's objective dominates, the bank risk might decrease with higher capital requirement. In contrast, a manager-driven bank is inclined to undertake more risk under tightened capital requirements because in such case, the manager's private benefit is larger with the increase in asset risk.

5. Global evidences

Reviews of theoretical models that explain the effect of capital requirements on bank risk behaviors show that the prediction of these models highly depends on model assumptions. This makes the results vary when certain assumptions are relaxed. Thus, a substantial effort has relied on empirical evidences to investigate the effect of the risk-based capital regulation on bank risk-taking. Most researches are carried out in the USA and the European countries, which are members of the Basel Committee. The results, however, differ across countries and time period (see **Table 2**).

Country	**Time period**	**Authors**	**Risk proxy**	**Capital regulation proxy**	**Effect of capital regulation on risk**
The USA	1990–1991	Jacques and Nigro [42]	Risk-weighted asset ratio	Gap approach	+ No effect for undercapitalized banks + Negative effect for well-capitalized banks
The USA	1990–1997	Aggarwal and Jacques [43]	Risk-weighted asset ratio	Dummy approach	+ Positive and no effect before 1993 + Negative effect during 1993–1996
The USA	2000–2005	Teply and Matejasák [44]	Risk-weighted asset ratio	Gap and Probabilistic approach	+ Negative effect for undercapitalized banks + No effect for well-capitalized banks
The USA	2003–2006	Abreu and Gulamhussen [45]	Risk-weighted asset ratio	Dummy and gap approach	Positive effect
Switzerland	1989–1995	Rime [46]	Risk-weighted asset ratio	Dummy and probabilistic approach	No effect
Germany	1994–2002	Heid et al. [47]	Risk-weighted asset ratio	Probabilistic and rolling-window approach	+ Negative effect for banks with low capital buffer + Positive effect for banks with high capital buffer
The USA, Canada, France, Italy, the UK, Japan	1988–1995	Van Roy [48]	Risk-weighted asset ratio	Probabilistic approach	No effect

Country	Time period	Authors	Risk proxy	Capital regulation proxy	Effect of capital regulation on risk
Italy	1994–2003	Cannata and Quagliariello [49]	+ Risk-weighted asset ratio + Bad debt ratio	Dummy approach	+ Positive effect for risk-weighted asset ratio + No effect for default risk
EU 15	2000–2005	Teply and Matejasák [44]	Risk-weighted asset ratio	Gap and Probabilistic approach	No effect
17 EU countries	1992–2006	Camara et al. [50]	+ Risk-weighted asset ratio + Nonperforming loan ratio + Default risk + Z-score	Dummy approach	+ Positive effect for highly capitalized and undercapitalized banks + Negative effect for adequately and strongly undercapitalized banks
G-10 Countries	1995–2005	Saadaoui [51]	Nonperforming loan ratio	Gap approach	No effect
India	1997–1998	Nachane and Saibal [52]	Risk-weighted asset ratio	Gap approach	No effect
China	2004–2006	Zhang et al. [32]	Risk-weighted asset ratio	Gap approach	No effect
Indonesia	2000–2005	Parinduri and Riyanto [53]	Risk-weighted asset ratio	Dummy and Probabilistic approach	Negative effect
42 Asian countries	1994–2008	Lee and Hsieh [54]	+ Variance of ROE + Loan loss reserve ratio	Capital regulatory index	+ Positive effect for variance of ROE + Negative effect for loan loss reserve ratio
Tunisia	1992–2005	Bouri [55]	+ Risk-weighted asset ratio + Loan loss provision ratio	Gap approach	+ Positive for risk-weighted asset ratio + No effect for loan loss provision ratio
Tunisia	2000–2013	Bouheni and Rachdi [56]	Risk-weighted asset ratio	Probabilistic approach	No effect
Brazil	2001–2009	Pereira and Saito [57]	Risk-weighted asset ratio	Dummy approach	Negative effect
Emerging markets	1996–2001	Godlewski [58]	Nonperforming loan ratio	Dummy and gap approach	+ No effect for dummy approach and for well-capitalized banks under gap approach + Negative effect for undercapitalized banks under gap approach
Emerging markets	1995–2005	Saadaoui [51]	Nonperforming loan ratio	Gap approach	Positive effect
Developing countries	1994–2002	Hussain and Hassan [59]	Risk-weighted asset ratio	Gap approach	Negative effect

Country	Time period	Authors	Risk proxy	Capital regulation proxy	Effect of capital regulation on risk
The US and non-US developed and developing countries	2003–2009	Lin et al. [60]	Nonperforming loan ratio	Probabilistic approach	Negative effect
GCC countries	1996–2011	Ghosh [61]	Z-score	Dummy approach	No effect
MENA countries	2004–2012	Bougatef and Mgadmi [62]	Loan loss provision ratio	Dummy approach	No effect
11 Dual banking countries	2006–2010	Alam [63]	Loan loss reserves ratio	Capital regulatory index	Negative effect
107 Countries	1999	Barth et al. [64]	Nonperforming loan ratio	Capital regulatory index	Negative effect

Notes: + Gap approach measures capital regulation as the distance of bank capital ratio from certain threshold, usually the minimum capital requirement.

+ Dummy approach assigns value 1 for banks whose capital ratios are less than certain threshold and 0 otherwise.
+ Probabilistic approach assigns value 1 for banks which are probably to be under regulatory pressure and 0 otherwise.

Table 2.
Global evidences about the effect of capital regulation on bank risk.

6. Conclusion

Efficient risk management is crucial for banks to ensure their profitability and maximize the shareholder's value. Over the past decades, the risk management practice has changed dramatically under the forces of the business environment and technology development. An important factor contributes to the way banks manage their risk is the regulation. This chapter shows how the regulators regulate bank risk with an emphasis on the risk-based capital regulation. It also reviews theoretical studies on how the risk-based capital regulation affects bank risk-taking. These studies explain the effect of capital regulation on bank risk considering different factors such as moral hazard, franchise value, capital buffer, and agency problem. The prediction of these studies is restricted by and depends on model assumptions. The chapter also provides empirical evidences from countries worldwide and shows that the effect of capital regulation on bank risk is not homogenous among countries.

DOI: http://dx.doi.org/10.5772/intechopen.79822

Author details

Nguyen Thi Thieu Quang[1] and Christopher Gan[2]*

1 Faculty of Banking, University of Economics, The University of Danang, Danang, Vietnam

2 Faculty of Agribusiness and Commerce, Lincoln University, Christchurch, New Zealand

*Address all correspondence to: christopher.gan@lincoln.ac.nz

References

[1] Basel Committee on Banking Supervision. Enhancements to the Basel II Framework. Basel, Switzerland: Bank for International Settlements; 2009

[2] Basel Committee on Banking Supervision. Principles for the Management of Credit Risk. Basel, Switzerland: Bank for International Settlements; 2000

[3] Basel Committee on Banking Supervision. Interest Rate Risk in the Banking Book. Basel, Switzerland: Bank for International Settlements; 2016

[4] Basel Committee on Banking Supervision. International Convergence of Capital Measurement and Capital Standards: A Revised Framework. Basel, Switzerland: Bank for International Settlements; 2006

[5] Basel Committee on Banking Supervision. Principles for Sound Liquidity Risk Management and Supervision. Basel, Switzerland: Bank for International Settlements; 2008

[6] Basel Committee on Banking Supervision. Principles for the Sound Management of Operational Risk. Basel, Switzerland: Bank for International Settlements; 2011

[7] Basel Committee on Banking Supervision. Corporate Governance Principles for Banks. Basel, Switzerland: Bank for International Settlements; 2015

[8] Basel Committee on Banking Supervision. Core Principles for Effective Banking Supervision. Basel, Switzerland: Bank for International Settlements; 2012

[9] Basel Committee on Banking Supervision. International Convergence of Capital Measurement and Capital Standards. Basel, Switzerland: Bank for International Settlements; 1988

[10] Basel Committee on Banking Supervision. Amendment to the Capital Accord to Incorporate Market Risks. Basel, Switzerland: Bank for International Settlements; 1996

[11] Basel Committee on Banking Supervision. International Convergence of Capital Measurement and Capital Standards: A Revised Framework. Basel, Switzerland: Bank for International Settlements; 2004

[12] Basel Committee on Banking Supervision. The Application of Basel II to Trading Activities and The Treatment of Double Default Effects. Basel, Switzerland: Bank for International Settlements; 2005

[13] Basel Committee on Banking Supervision. Basel III: A Global Regulatory Framework for More Resilient Banks and Banking Systems. Basel, Switzerland: Bank for International Settlements; 2010

[14] Basel Committee on Banking Supervision. Basel III: The Liquidity Coverage Ratio and Liquidity Risk Monitoring Tools. Basel, Switzerland: Bank for International Settlements; 2013

[15] Basel Committee on Banking Supervision. Basel III: The Net Stable Funding Ratio. Basel, Switzerland: Bank for International Settlements; 2014

[16] Basel Committee on Banking Supervision. Basel III: Finalising Post-Crisis Reforms. Basel, Switzerland: Bank for International Settlements; 2017

[17] Mishkin FS. Financial Markets and Institutions. In: Eakins SG, editor. 8th ed. Boston: Pearson; 2015

[18] Kim D, Santomero AM. Risk in banking and capital regulation. The Journal of Finance. 1988;**43**(5):1219-1233

[19] Flannery MJ. Capital regulation and insured banks choice of individual loan default risks. Journal of Monetary Economics. 1989;**24**(2):235-258

[20] Furlong FT, Keeley MC. Capital regulation and bank risk-taking: A note. Journal of Banking & Finance. 1989;**13**(6):883-891

[21] Gjerde Ø, Semmen K. Risk-based capital requirements and bank portfolio risk. Journal of Banking & Finance. 1995;**19**(7):1159-1173

[22] Zweifel P, Pfaff D, Kühn J. A simple model of bank behaviour—With implications for solvency regulation. Studies in Microeconomics. 2015;**3**(1):49-68

[23] Gennotte G, Pyle D. Capital controls and bank risk. Journal of Banking & Finance. 1991;**15**(4):805-824

[24] Rochet J-C. Capital requirements and the behaviour of commercial banks. European Economic Review. 1992;**36**(5):1137-1170

[25] Calem P, Rob R. The impact of capital-based regulation on bank risk-taking. Journal of Financial Intermediation. 1999;**8**(4):317-352

[26] Blum J. Do capital adequacy requirements reduce risks in banking? Journal of Banking & Finance. 1999;**23**(5):755-771

[27] Silva N. Capital regulation and bank risk taking: Completing Blum's picture. Documentos de Trabajo (Banco Central de Chile). 2007;(416):1

[28] Northcott CA. Competition in banking: A review of the literature. Working Paper. Bank of Canada; 2004

[29] Demsetz RS, Saidenberg MR, Strahan PE. Banks with something to lose: The disciplinary role of franchise value. Economic Policy Review—Federal Reserve Bank of New York. 1996;**2**(2):1-14

[30] Hellmann TF, Murdock KC, Stiglitz JE. Liberalization, moral hazard in banking, and prudential regulation: Are capital requirements enough? The American Economic Review. 2000;**90**(1):147-165

[31] Repullo R. Capital requirements, market power, and risk-taking in banking. Journal of Financial Intermediation. 2004;**13**(2):156-182

[32] Zhang Z-Y, Wu J, Liu Q-F. Impacts of capital adequacy regulation on risk-taking behaviors of banking. Systems Engineering—Theory & Practice. 2008;**28**(8):183-189

[33] Behr P, Schmidt RH, Xie R. Market structure, capital regulation and bank risk taking. Journal of Financial Services Research. 2010;**37**(2):131-158

[34] Buser SA, Chen AH, Kane EJ. Federal deposit insurance, regulatory policy, and optimal bank capital. The Journal of Finance. 1981;**36**(1):51-60

[35] Milne A, Whalley AE. Bank Capital Regulation and Incentives for Risk-Taking. Cass Business School Research Paper; 2001

[36] Milne A. Bank capital regulation as an incentive mechanism: Implications for portfolio choice. Journal of Banking & Finance. 2002;**26**(1):1-23

[37] Berger AN, Herring RJ, Szegö GP. The role of capital in financial institutions. Journal of Banking & Finance. 1995;**19**(3-4):393-430

[38] Jensen MC, Meckling WH. Theory of the firm: Managerial behavior, agency costs and ownership structure. Journal of Financial Economics. 1976;**3**(4):305-360

[39] Besanko D, Kanatas G. The regulation of bank capital: Do capital standards promote bank safety? Journal of Financial Intermediation. 1996;**5**(2):160-183

[40] Gale D. Capital regulation and risk sharing. International Journal of Central Banking. 2010;**6**(4):187-204

[41] Jeitschko TD, Jeung SD. Incentives for risk-taking in banking—A unified approach. Journal of Banking & Finance. 2005;**29**(3):759-777

[42] Jacques K, Nigro P. Risk-based capital, portfolio risk, and bank capital: A simultaneous equations approach. Journal of Economics and Business. 1997;**49**(6):533-547

[43] Aggarwal R, Jacques KT. The impact of FDICIA and prompt corrective action on bank capital and risk: Estimates using a simultaneous equations model. Journal of Banking & Finance. 2001;**25**(6):1139-1160

[44] Teply P, Matejasák M. Regulation of Bank Capital and Behavior of Banks: Assessing the US and the EU-15 Region Banks in the 2000-2005 Period. St. Louis: Federal Reserve Bank of St Louis; 2007

[45] Abreu F, Gulamhussen M. The Relationship between Capital Requirements and Bank Behavior: A Revision in the Light of Basel II. Aarhus: European Financial Management Association; 2010

[46] Rime B. Capital requirements and bank behaviour: Empirical evidence for Switzerland. Journal of Banking & Finance. 2001;**25**(4):789-805

[47] Heid F, Porath D, Stolz S. Does Capital Regulation Matter for Bank Behavior? Evidence for German Savings Banks. St. Louis: Federal Reserve Bank of St Louis; 2003

[48] Van Roy P. Capital requirements and bank behavior in the early 1990s: Cross country evidence. International Journal of Central Banking. 2008;**4**(3):29-60

[49] Cannata F, Quagliariello M. Capital and risk in Italian banks: A simultaneous equation approach. Journal of Banking Regulation. 2006;**7**(3-4):283-297

[50] Camara B, Lepetit L, Tarazi A. Ex ante capital position, changes in the different components of regulatory capital and bank risk. Applied Economics. 2013;**45**(34):4831-4856

[51] Saadaoui Z. Risk-based capital standards and bank behaviour in emerging and developed countries. Journal of Banking Regulation. 2011;**12**(2):180-191

[52] Nachane DM, Saibal G. Risk-based standards, portfolio risk and bank capital: An econometric study. Economic and Political Weekly. 2001;**36**(10):871-876

[53] Parinduri RA, Riyanto YE. Do banks respond to capital requirements? Evidence from Indonesia. Applied Financial Economics. 2011;**21**(9):651-663

[54] Lee C-C, Hsieh M-F. The impact of bank capital on profitability and risk in Asian banking. Journal of International Money and Finance. 2013;**32**:251-281

[55] Bouri A, Ben Hmida A. Capital and risk taking of banks under regulation: A simultaneous equations approach in the Tunisian context. In: Proposition pour le Sixième Congrès Intrenational de l'AFFI: Finance d'enterprise et finance de marchè: quelles complèmentarités? 2006

[56] Bouheni FB, Rachdi H. Bank capital adequacy requirements and risk-taking behavior in Tunisia: A

simultaneous equations framework. Journal of Applied Business Research. 2015;**31**(1):231

[57] Pereira JAM, Saito R. Coordination of capital buffer and risk profile under supervision of Central Bank (Coordenação entre capital buffer e perfil de risco sob supervisão do Banco Central). Revista Brasileira de Finanças. 2015;**13**(1):73-101

[58] Godlewski CJ. Bank capital and credit risk taking in emerging market economies. Journal of Banking Regulation. 2005;**6**(2):128-145

[59] Hussain ME, Hassan MK. Basel Capital Requirements and Bank Credit Risk Taking in Developing Countries. Department of Economics and Finance Working Papers, 1991-2006. Paper 34. 2005

[60] Lin SL, Hwang D-Y, Wang KL, Xie ZW. Banking capital and risk-taking adjustment under capital regulation: The role of financial freedom, concentration and governance control. International Journal of Management, Economics and Social Sciences. 2013;**2**(2):99-128

[61] Ghosh S. Risk, capital and financial crisis: Evidence for GCC banks. Borsa Istanbul Review. 2014;**14**(3):145-157

[62] Bougatef K, Mgadmi N. The impact of prudential regulation on bank capital and risk-taking: The case of MENA countries. The Spanish Review of Financial Economics. 2016;**14**(2):51-56

[63] Alam N. Regulations and bank risk taking in dual banking countries. Journal of Banking Regulation. 2014;**15**(2):105-116

[64] Barth JR, Caprio G, Levine R. Bank regulation and supervision: What works best? Journal of Financial Intermediation. 2004;**13**(2):205-248

Chapter 5

Risk Management Practices Adopted by European Financial Firms with a Mediterranean Connection

Simon Grima and Frank Bezzina

Abstract

Following the economic and financial crises, any activity involving internal controls, especially risk management, has been given more attention. With this study, we aim to contribute further to the existing literature on risk management by looking at practices adopted by financial services firms licenced in Europe with a Mediterranean connection. We used parts of a questionnaire adopted by two of the authors in another study on risk management practices adopted by Maltese financial services firms and sent it to prospective candidates who work closely within risk management, to collect our data. This resulted in 1635 participants. This data was used to (1) bring to light the mechanisms and strategies used in risk management by these organisations to maximise their opportunities, manage their risks, and maintain stability in their financials. Also, (2) we check if this is perceived as contributing to 'principled performance'. Finally, (3) we examine the extent to which risk management capabilities offer a competitive advantage to these firms. Our findings evidence that the objective by EMP and the EU, that is to ensure that members operate 'on the same level playing field' within risk management, in financial services of firms with a Euro-Mediterranean connection, has been achieved.

Keywords: risk management, financial services industry, risk management frameworks, Euro-Mediterranean, principled performance

JEL code

G2, G3

1. Introduction

Although risks have been present since the beginning of mankind, explicit attention to them has differed over the passage of time. Early civilisations attributed unexpected events to the gods. This made it pointless for mankind to intervene and manage. This continued till today with some tribes in some parts of the world and was even the case until the middle ages when Christianity was strong in the current Western Europe. The word "risk" itself is derived from ancient Arabic word "rizq" and used till today in the Maltese Language, translated to mean prosperity granted by God (Allah) to a person. However, during the Renaissance, in Europe this was given the meaning of an uncertain loss or danger Doff, [1].

Following the economic and financial crises of this century, any activity involving internal controls, especially risk management has been given more attention and importance. This, as noted in the World Economic Forum [2], was due mainly to the successful results of effective risk management during periods of global economic turbulence [8]. In fact, as Ghoshal [3] highlighted, one of the main objectives of any organisation is to manage their risk.

However, the treatment and understanding of risk and as a consequence its management, varies both in literature and in practice. Moreover, as March and Shapira [4] note, the strategic management field does not provide us with one specific accepted definition of risk and highlight that most managers view risk as a negative outcome.

Hillson [5] defines risk as an "uncertainty that matters because it can affect one or more objectives". Also, literature by [6], show that one needs to distinguish between the known, unknown and unknowable uncertainties before defining what constitutes a risk and as a consequence managing it under the risk management process. Unknowable uncertainty is when the missing information is unavailable to all known uncertainty is when the probability is an objective chance and is generally agreed upon and unknown uncertainty is when the probability may be or is known by somebody [6].

The strategy of any organisation has to deal with the alignment to its uncertain environment and to rebalance its strategic choices to determine the exposure to this uncertain environment, which impacts performance. To this effect various studies have focussed on understanding the risk management discipline and practices of firms in specific activities, areas and countries. Moreover, the effectiveness and efficiency of appropriate practices in risk management is critical for the continued existence, industry profitability and for the continual development and growth of the whole economy. It is imperative that all organisations adopt good quality practices and measures when managing risks [7].

With this study, we aim to contribute further to the existing literature on the risk management by looking at practices adopted by financial services firms licenced in Europe with a Mediterranean connection, specifically Cyprus, France, Italy, Spain, Croatia, Greece, and Slovenia extending and comparing to the work of Bezzina et al. [7] on Malta. We chose members, which although, have inherent country and cultural diversity and are joined by their geographical border with the Mediterranean Sea, aim for a level regulatory and economic playing field through their union in the European Union (EU) and the Euro-Mediterranean partnership [8].

When dealing with financial services firms, in the EU, this regulatory level playing field is much more pronounced, since financial firms are required to abide by common directives such as the Capital Requirement Directive (CRD) in banks and investment firms, Solvency II (SII) in Insurance firms and other soft laws. This is likely to make the sample more representative and the empirical results more generalisable. It will also shed light on whether European Union within the Euro-Mediterranean region and the Euro-Mediterranean Partnership (EMP), has brought these countries closer together in practices, specifically when dealing with risk and its management.

We use part of the questionnaire adopted by Bezzina et al. [9] in their paper on risk management practices adopted by Maltese financial services firms, to collect our data and (1) bring to light the mechanisms and strategies used in risk management by these organisations to maximise their opportunities, manage their risks, and maintain stability in their financials. Also, (2) we check if this is perceived as contributing to 'principled performance' (defined in the chapter in Section 2.2). Finally (3) we examine the extent to which risk management capabilities offer a competitive advantage to these firms.

2. Literature

We can cite various studies dealing with risk management practices in different areas, industries, regions and countries. For example, a study on risk management practices carried out on the Ghanaian insurance industry by [7] revealed that companies insuring life, different from companies insuring non-life, have their risk appetite levels statements recorded. This enables the identification of those risks to on-board and those ones to transfer. Moreover, they exposed that the industry lacks adequate skilled personnel and risk management is reactive as a response to regulatory directives. Other surveys carried out about the UK insurance industry showed that the response by most insurance firms to risk management regulations was perfunctory, rather than being seen as good business practice [10].

Another study by [11] on risk management practices of German firms revealed that participants showed no difficulty in developing a risk management system and rated business survival as the top risk management goal. Moreover, they showed that respondents are more risk-neutral than risk-averse for financial risks, and that 88 percent use derivatives.

Bankers operating in Barbados perceived risk management as critical to the performance of their banks; with operational risk, credit risk, country/sovereign risk, market risk and interest rate risk being their greatest exposures [12], while those operating in Bahrain show a clear understanding of both risk and risk management and have efficient risk assessment analysis, risk identification processes, credit risk analysis, risk monitoring and risk management practices with credit, liquidity and operational risk being the most prominent risks faced by both conventional and Islamic banks [13].

A study on Islamic banks in Pakistan showed that they are efficient in managing their risks. Revealing that the most influencing variables in the risk management process were that of understanding risk and risk management, risk monitoring and credit risk analysis [14]. On the other hand, Hassan [15], found that the Islamic banks in Brunei Darussalam consider foreign-exchange risk, followed by credit risk and then operating risk, as the 3 most important risks. He also noted that Islamic banks are very efficient mainly in risk identification, assessment and analysis.

A further study by Sifumba et al. [16] revealed that manufacturing SMEs personnel in Cape Town are not aware of the elements that make risk management effective. While in Malta, Bezzina et al. [8], found that financial firms have a strong culture of efficient and effective risk management practices that add value and are linked to well-defined objectives with corporate social responsibility embedded within the organisations' risk management corporate strategies and corporate culture. Miloš Sprčić et al. [17], in a study on Croatian companies, find that the risk management system development is dependent only on value of the growth options and the size of the company.

2.1 Risk management strategies and mechanisms

Any organisation's strategy needs to deal with an uncertain environment. Therefore, organisational strategic choices will determine the organisation's exposure to an uncertain environmental and constituents that impact their performance. "Exposure" defined as the sensitivity of an organisation's cash flows to changes in interrelated uncertain variables. The emphasis of organisation on specific particular (particularist view) rather than multidimensional uncertainties is a significant shortcoming. The former view of isolating specific uncertainties, excludes other interrelated uncertain variables. In fact, literature in financial services emphasises uncertainties for which hedging or insurance instruments can be designed to manage organisation exposures, however omitting some uncertainties that are encountered in the overall management strategic decisions. The alternative view is where management takes a general approach to risk and gives explicit consideration to numerous uncertainties (integrated risk management perspective) [18].

Das and Teng [19], build on the latter and suggests that to effectively manage risks and reduce unwanted risks, organisations need to examine the inter-relationship between trust, control and risk using an integrated framework which examines the inter-relationship between the three constructs. They note that firms need to manage their risks by determining the conjoint roles of these constructs in the context of their objectives and strategies.

It has therefore always been a must for every leading firm to ensure that the process of identifying risk and managing it, is an explicit part of the strategic plan, and that there is a buy-in from all levels of their organisation. Risk management should be seen as a systematic effort that is pervasive through all operating units, be it in the front, mid or back office, right in line with growth areas targeted for investments or any critical support functions. Risk management must matter to the organisation and to the person whose occupation and responsibility is defined by it [20].

The risk manager or officer is responsible to initiate the process of determining the risks faced by the company, based on the strategy, determine the mandatory and voluntary barriers and put in place a risk management strategy to achieve objectives with the least of problems. That is the objective risk assessment process which depends on the organisation, and the plan and tactics to arrive at that objective [21].

Stulz [22] offers us theoretical evidence showing that risk management practice within firms is limited. Marshall and Heffes [23] report that only 11 percent of "more than 90 percent of the executives who say they are building or want to build enterprise risk management (ERM) processes into their organization report they have completed their implementation. The survey results indicate that more than two-thirds of both boards of directors and senior management staff consider risk management to be an important responsibility". COSO's recent survey [24] findings show unsatisfactory results for the implementation of ERM showing that "60 percent of respondents say their risk tracking is mostly informal and ad hoc or only tracked within individual silos or categories as opposed to enterprise-wide."

2.2 Risk management and principled performance

As explained in Bezzina et al. [8] we again adopted the Open Compliance and Ethics Group's (OCEG) standard's concept of integrating internal controls "(the Governance, Risk Management and Compliance (GRC) capability model) into one main function [24]. This as suggested by these authors and OCEG, helps to "improve quality and performance, by providing tools that can measure and enhance corporate culture within an integrated environment." This structure is said to be the main determinant of the achievement of 'Principled Performance' as defined by OCEG. That is "reliable achievement of objectives while addressing uncertainty and acting with integrity." [21].

"OCEG in their definition of 'Principled Performance' emphasises the unambiguous articulation of a firm's objectives in financial and non-financial form. It outlines the methods and boundaries that would be adhered to while achieving the set targets." They continued to note that 'Principled Performance' in a financial firm can be achieved with clearly defined objectives, goals, values and a transparent, effective flexible mechanism, which enables continuous improvement to address risks and vulnerabilities within established boundaries [25].

Mitchell [25] continues by highlighting that, mainly if the existing structure offers a competitive advantage, GRC requires function integration without the need for operations consolidation. One can replicate the strengths of approaches, communication, technology used and reporting integration to the whole business to benefits from reduced errors, better information quality, and reduced costs. The GRC 360 Capability Model, 2009 specifies that, while culture, structure and the organisation play an essential role in the overall performance of a company; people, process and technology are crucial for principled performance.

2.3 Risk management abilities and competitive advantage

Creativity is lost if we only think of risk management as a way to minimise risk. We need to take risks and if and when they go in some unwanted unpredictable path, we need to be able to respond to them [26]. Kannan and Thangavel [27] note that every major advance in human civilization was possible because someone was willing to take a risk and challenge the status quo.

Enterprise risk management (ERM) promotes risk management as a more strategic responsibility and emphasises that if effectively implemented it can create a long term competitive advantage [28]. However, Slywotzky and Drzik [29] suggest, that many companies still treat ERM as an extension of their internal control processes, while only a few companies, use their risk management abilities as a source of competitive advantage. In fact these companies go beyond internal controls and cost-controlling (defensive and reactive approaches), taking a more aggressive and proactive stance towards risk. These have understood that managing risk is a source of leverage to gain competitive advantage [30].

Ehsan [30], limited risks faced by a company, to two major types: rewarded and unrewarded risks, and continues to note that the way through which capabilities of risk management can increase competitive advantage depends mainly on the type of risk exposure the company has. Rewarded risks are those risks that are expected to gain us some type of benefit, that is, risks taken to create value and are consequences of our decisions. Unrewarded risks usually brought about by external forces, such as natural disasters, industrial accidents, theft, pandemics, etc. which have no potential value in them. The ability to effectively deal with these risks has an important impact on the company's performance and thereby its competitive advantage.

In his seminal book, Porter [31] argues that "there are two major ways that a company can gain competitive advantage over its competitors: cost advantage, and differentiation". Risk management capabilities can help to affect the company's costs and the value it creates for stakeholders. Moreover, in theory, since risk management is a proactive activity, it can help create preparedness and advanced warnings for disruptions (i.e., to ensure business continuity). This differentiates these companies from their competitors giving them a competitive advantage [30].

3. Method

A questionnaire adopted from a previous study by Bezzina et al. [8] to determine the risk management practices by Maltese financial firms, was used to extend this study to other Euro-Mediterranean countries. This questionnaire was administered to persons working in, or with a connection to the field of risk management within the financial services industry. Participants for the questionnaire were recruited with the help of one of the authors who is an active participant in European risk management associations. The survey was administered using an online questionnaire which was opened in January 2017 and closed in November 2017. In the introduction page we outlined our aims and objectives, while in the next four sections we posed closed-ended statements, which related to four main themes: (i) strategies and mechanisms adopted in risk management; (ii) the perceived purpose, scope and benefits of risk management; (iii) risk management and competitive advantage; and (iv) CSR influences on the corporate risk management strategies. The participants were asked to choose from a five-point Likert scale mainly ranging from 'strongly disagree' (coded as '1') to 'strongly agree' (coded as '5'), and some others ranging from 'very unimportant' (coded as '1') to 'very important' (coded as '5'). The final section (Section 5) was dedicated to the collection of the demographic data about the participant and their organisations. This data was collected in the form of labels or a scale and presented in aggregate, so as not to enable identification of the organisation or the participant. The responses (1635 respondents) were then subjected to statistical analysis using SPSS. When summarising the data, the median and interquartile range (lower quartile to upper quartile) were used for the ordinal scales while the mean and standard deviation were used with the interval/ratio scales. To test for differences in mean ranks, the Friedman test (a non-parametric alternative to one-way ANOVA) was used. Participants were guaranteed that their identity and that of the firm they are representing will be maintained anonymous.

4. Research questions

As noted above, being an extension of a previous study by Bezzina et al. [8], and since we adopted the same questionnaire, we also maintained the same research questions and the new responses were used to investigate and compare to the findings in that study.

4.1 RQ1

What are the risk management strategies and mechanisms adopted by financial services firms within countries with a Euro-Mediterranean connection in order to manage their risks, strengthen their opportunities and retain financial stability?

4.2 RQ2

Do the financial services firms in countries with a Euro-Mediterranean connection perceive risk management as just an authority imposed obligatory requirement or do they see it as critical for the achievement of 'principled performance'?

4.3 RQ3

Do risk management abilities offer a competitive advantage to financial services firms in countries with a Euro-Mediterranean connection?

5. Results

5.1 RQ1

The respondents reported (on average), that they strongly agree their institution has a strategic risk management plan in place (Md = 5, IQR = 4–5). Furthermore, they agreed (Md = 4), that they have systems in place to strengthen the risk management process (see **Table 1**).

Furthermore, we asked the respondents to rate their level of agreement with seven statements related to the scope of their institution's strategic risk management plan is. **Table 2** shows that there were significant differences in mean ranks based on the Friedman test, although they strongly agreed (Md = 5) or agreed (Md = 4) with all the statements.

We then delved into the quality requirements of risk management, the procedures/processes/policies of risk management, and the risk culture adopted by the financial services institutions. Details of the items used and a summary of statistical output are provided in **Table 3**. The responses exhibit empirical evidence of a strong risk management within the institutions investigated.

Finally, we wanted to know how important each of 12 established frameworks were for the institutions when implementing risk management. **Table 4** shows that four frameworks were overall rated as 'important', (Basel Accords, COSO 2, IAS and Interest Rate Risk Management), and the remaining eight as 'neither important nor unimportant'.

5.2 RQ2

The respondents reported (on average) that risk management practices play a vital role in their institutions (M = 4.00; SD = 0.51) and have a positive perception of risk management practices in achieving principled performance (M = 4.10, 0.64). **Table 5** provides a summary of the responses and statistical output pertaining to the individual items that make up these two constructs.

Furthermore, they agreed (Md = 4) that they give sufficient attention to all the risks that we outlined when designing strategies and objectives, bar 'health and safety' (Md = 3). **Table 6** exhibits the 14 risks in order of decreasing attention as rated by the respondents.

5.3 RQ3

In this research question, we wanted to better understand the capabilities of risk management practices in achieving competitive advantage. We first sought to determine the institutions' intention behind the risk management strategy for the financial services (see **Table 7**).

The strategic risk management plan	Median	IQR
Is clearly communicated and understood	4	3–5
Is a contribution of all team members	4	3–5
Is a responsibility of top management	4	4–5
Is in sync with individual risk management plans	4	4–5

N = 1635; scales are ordinal and range from strongly disagree ('1') to strongly agree ('5').

Table 1.
Summary statistics.

The scope of the strategic risk management plan	Median	IQR	Mean rank
Is to provide a framework for the risk management process of identification, monitoring, control and decision	5	4–5	4.76
Is to provide the appropriate setup to enable risk assessments in terms of costs and benefits of identified risks	5	4–5	4.73
Is to help maintain stability in financials	4	4–5	3.94
Is to allow for innovation to maximise opportunities and cost reduction	4	4–5	3.68
Is to provide a framework with roles and responsibilities to enable better risk identification	4	4–5	3.66
Is to record declared aims and objectives and ensure a systematic identification of risks relating to each	4	4–5	3.62
Is to provide a defined structure to sustain business growth and continued profitability within objectives, appetite and tolerance.	4	4–5	3.60

Friedman test: $\chi^2(6) = 1777.37$, $p \leq 0.001$; N = 1635; scales are ordinal and range from strongly disagree ('1') to strongly agree ('5').

Table 2.
Summary statistics and Friedman test output.

Statement	Median	IQR	Mean rank
In our institution...[a]			
The risk manager is an active member of the risk management committee	4	4–5	2.76
The risk management committee members communicate the risk appetite and tolerance of the firm	4	4–4	2.52
The risk management committee members are Knowledgeable	4	4–4	2.37
The risk manager makes use of bottom up methodologies in developing the strategic risk management plan	4	4–4	2.36
Which of the following initiatives are embedded within the firm's risk management strategy?[b]			
Risk reporting and information systems	5	4–5	3.67
Enterprise risk management practices	5	4–5	3.64
Ongoing improvements in risk management practices	5	4–5	3.62
Risk measurement and monitoring in non-financial terms	5	4–5	3.59
Risk measurement and monitoring practices in financial terms	5	4–5	3.47
Identification and quantification of risks and controls	4	4–5	3.01
To what extent does your institution map its risks (identification, description and prioritisation)?[c]			
Top down approach and bottom up approach	5	4–5	3.53
On a global corporate level only (strategic, financial and operational)	4	4–5	3.43
Risks are managed at group level or are silo based*	4	3–5	2.92
Only for certain business units/areas*	4	2–5	2.56
Only for certain categories of risks*	4	2–5	2.56

N = 1635; scales are ordinal and range from strongly disagree ('1') to strongly agree ('5'); Friedman tests: $\chi^2(3) = 391.16$, $p \leq 0.001$.

[a] $\chi^2(6) = 749.48$, $p \leq 0.001$.

[b] $\chi^2(5) = 749.48$, $p \leq 0.001$.

[c] $\chi2(4) = 1028.64$, $p \leq 0.001$.
**Reverse coded.*

Table 3.
Risk management quality requirements, risk management procedures/process/policies and risk culture: summary statistics and Friedman test output.

Our institution makes use of the following frameworks when implementing risk management	Median	IQR (range)	Mean rank
Basel Accords	4	4–5	9.14
COSO 2	4	3–5	8.72
International Accounting Standards (IAS)	4	3–4	7.91
ISO 31000	4	3–5	7.79
Interest Rate Risk Management (e.g., duration or gap analysis)	4	3–4	6.70
FERMA	3	3–4	6.53
Cobit	3	3–4	6.47
National Risk Management Standards (NRMS)	3	3–3	5.55
Value at Risk (VAR)	3	3–4	5.46
Prince 2	3	3–4	5.37
AIRMIC	3	3–3	4.20
ITIL	3	3–3	4.17

N = 1635; scales are ordinal and range from very unimportant ('1') to very important ('5'); Friedman test: $\chi^2(11) = 5906.88, p \leq 0.001$.

Table 4.
Risk management frameworks: summary statistics and Friedman test output.

	Median	IQR
Risk management practices play a vital role in ensuring that our institution		
Clearly defines its goals and values	4	4–4
Outlines how these goals are achieved	4	4–4
Identifies and demonstrates how risks and vulnerabilities would be addressed	4	4–4
Allows for transparency with stakeholders	4	3–4
Implements an effective mechanism for change, enabling continuous improvement to achieve the desired outcomes	4	4–5
Perceived purpose of risk management practices		
Critical factor in achieving principled performance	4	4–5
Vital to the performance and success of our institution's objectives	4	3–5
No link between principled performance and RM practices*	4	4–5
Principled performance does not form part of our institution's risk management practices benefit realisation plans*	4	4–5
Puts a strain on resource effort for compliance's purposes without providing added value*	4	3–5

*N = 1635; scales are ordinal and range from strongly disagree ('1') to strongly agree ('5'). *Reverse coded.*

Table 5.
Role and perceived purpose of risk management practices: summary statistics.

We then wanted to examine the extent to which continuous risk impact assessments strengthen the competitive advantage in each of 8 factors. **Table 8** shows that the respondents agreed (Md = 4) with all the factors bar political and legal factors (Md = 3).

Furthermore, we examined the benefits risk management capabilities provide to institutions. These respondents agreed (Md = 4) that risk management infuses a

Our institution gives sufficient attention to the following risks when designing strategies and objectives	Median	IQR	Mean rank
Credit risk	4	4–5	9.41
Financial risk	4	4–5	9.01
Liquidity risk	4	4–5	8.96
Fraud risk	4	4–5	8.52
Operational risk	4	4–5	8.48
Strategic risk	4	3–5	8.23
Market risk	4	4–4	7.73
Reputation risk	4	4–5	7.64
External risk	4	4–5	7.46
Corporate governance risk	4	3–4	6.52
Legal/ethical risk	4	3–4	6.33
Administrative risk	4	3–4	6.00
Information risk	4	3–4	5.75
Health and safety risks	3	3–4	5.28

N = 1635; scales are ordinal and range from very low ('1') to very high ('5'). Friedman test: $\chi^2(13) = 3839.11, p \leq 0.001$.

Table 6.
Risks when designing strategies and objectives: summary statistics and Friedman test output.

The risk management strategy was implemented...	Median	IQR	Mean rank
To abide by legal, regulatory or compliance requirements	5	3–5	6.11
To formally define the institution's risk appetite	5	3–5	5.92
To formalise the governance structure	4	3–5	5.84
For catastrophic events or major crises (reaction to unexpected losses)	4	3–5	5.81
For corporate social responsibility	4	3–5	5.51
Due to pressure from analysts and/or rating agencies	4	3–5	5.32
Due to pressure from the market (e.g., competitors, suppliers, etc.)	4	3–5	5.28
To instil a consistent strong risk culture focussed on optimising understood risk return trade-offs within the defined risk strategy	4	3–5	5.13
To ensure full transparency across all risks and across the organisation	4	3–5	5.08
For competitive advantage	4	3–5	5.01

Friedman test: $\chi^2(9) = 1057.85, p \leq 0.001; N = 1635$*; scales are ordinal and range from strongly disagree ('1') to strongly agree ('5').*

Table 7.
Intention behind the risk management strategy: summary statistics and Friedman test output.

risk culture in the institution (IQR = 3–5), sustains future profitability (IQR = 4–5), provides visibility of economic and financial environment (IQR = 3–5) as well as long term profitable growth (IQR = 3–4) and provides competitive advantage (IQR = 3–5). Furthermore, we asked the respondents to rate their level of agreement with six factors aimed strengthening core risk management functions. The findings are exhibited in **Table 9**.

Our institution's risk management strategy requires that continuous risk impact assessments are conducted in order to strengthen the competitive advantage in:	Median	IQR	Mean rank
Financial capabilities	4	4–5	5.79
Economic factors	4	4–5	5.64
Marketing capabilities	4	3–5	4.40
Competitive factors	4	3–5	4.38
Cultural and societal factors	4	3–5	4.18
Technology	4	3–4	3.97
Human resource capabilities	4	3–5	3.74
Political and legal factors	3	3–4	3.90

N = 1635; scales are ordinal and range from strongly disagree ('1') to strongly ('5'); Friedman test: $\chi^2(7) = 2095.78$, $p \leq 0.001$.

Table 8.
Factors strengthening competitive advantage: summary statistics and Friedman test output.

In order to strengthen the core risk management functions, our executives seek to...	Median	IQR	Mean rank
Carry out continuous risk analysis of its credit portfolio	4	3–5	3.74
Adjust credit policies and revise mandates and incentive systems	4	3–5	3.73
Strengthen the internal information markets to make information available to decision makers on credit and sources of finance	4	3–5	3.65
Continuously strengthen internal capital efficiency and capital planning for the coming years to reflect potential market scenarios	4	3–5	3.45
Carry out strategic re-adjustment of liquidity intensive businesses	4	3–5	3.26
Refine the risk management tools to optimise usage of liquidity and improve transparency	4	3–4	3.17

N = 1635; scales are ordinal and range from strongly disagree ('1') to strongly agree ('5'); Friedman test: $\chi^2(5) = 483.38$, $p \leq 0.001$.

Table 9.
Strengthening core risk management functions: summary statistics and Friedman test output.

6. Conclusion

Our findings evidence that although authors such as Youngs [9], show strong scepticisms on the works and challenges of the EMP and the EU legislation; mainly to ensure that members operate on the same level playing field; within risk management in financial services of firms with a Euro-Mediterranean connection, this objective has been achieved. In fact, results show that similarly to the findings by Bezzina et al. [8] on Maltese financial services firms, personnel working or are involved in/with risk management of financial services firms with their head offices operating from Cyprus, France, Italy, Spain, Croatia, Greece, and Slovenia report that they have a strategic risk management plan in place with systems to enable the strengthening of their risk management processes to reach clearly identified objectives. They note various reasons that have helped to ensure this, with the strongest reasons being that of abiding to legal, regulatory and compliance requirements and the need to have a framework for systematic risk identification, mitigation, management, monitoring and control.

Findings, also show that the risk manager in these firms, similar to that of Maltese financial services forms, is highly active and involved, very knowledgeable and uses both top-down and bottom-up approaches to communicate the risk appetite of the company. This is facilitated by the fact that the quality and importance of risk management is embedded within their risk management strategy and seen as part of the firms' growth road map and a way to meet objectives. Moreover, in carrying out and designing their risk management strategy and processes these institutions tend to favour the use of frameworks/recommendations with the most followed being that provided by the Basel Accords. However, although, they give attention to practically all known risks identified, they are neutral on 'health and safety' issues, maybe because this might fall out of the competence of the respondents.

Finally, findings show that risk management practices play a vital role in ensuring that institutions reach their objectives (principled performance), add value and create a competitive advantage. This , with these practices, goals and values, is being clearly recorded and communicated; the roadmap to successfully reaching objectives is transparent and clear, enabling appropriate, identification of risks, growth, profitability, flexibility for improvement and change and quick response to uncertainties.

Author details

Simon Grima[1*] and Frank Bezzina[2]

1 Insurance Department, Faculty of Economics, Management and Accountancy, University of Malta, Malta

2 Management Department, Faculty of Economics, Management and Accountancy, University of Malta, Malta

*Address all correspondence to: simon.grima@um.edu.mt

References

[1] Doff R. Risk Management for Insurers, Risk Control, Economic Capital and Slvency II (3rd ed). Risk Books; 2015. pp. 11-13

[2] Schwab K, Sala-I-Martin X. The Global Competitiveness Report 2009-2010. In: Schwab K, editor. World Economic Forum. 2009. http://www3.weforum.org/docs/WEF_GlobalCompetitivenessReport_2009-10.pdf [Accessed 21st August, 2018]

[3] Ghoshal S. Global strategy: An organizing framework. Strategic Management Journal. 1987;**8**:425-440

[4] March JG, Shapira Z. Managerial perspectives on risk and risk taking. Management Science. 1987;**33**:1404-1418

[5] Hillson DA. What is risk? Towards a common definition. Journal of the UK Institute of Risk Management. 2002;**2002**:11-12

[6] Chua CC, Sarin RK. Known, unknown, and unknowable uncertainties. Theory and Decision. 2002;**52**(2):127-138. DOI: 10.1023/A:1015544715608 [Accessed 23rd June, 2018]

[7] Akoteyl JO, Abor J. Risk management in the Ghanaian insurance industry. Qualitative Research in Financial Markets. 2013;**5**(1):26-42. DOI: 10.1108/17554171311308940

[8] Bezzina F, Grima S, Mamo J. Risk management practices adopted by financial firms in Malta. Managerial Finance. 2014;**40**(6):587-612

[9] Youngs R. 20 years of the Euro-Mediterranean Partnership. Mediterranean Politics. Carnegie Europe. 2015. https://carnegieeurope.eu/2015/05/18/20-years-of-euro-mediterranean-partnership-pub-60337 [Accessed 22nd June, 2018]

[10] Dowd K, Bartlett DL, Chaplin M, Kelliher P, O'Brien C. Risk Management in the UK insurance industry: The changing state of practice. In: CRIS Discussion Paper Seroes – 2007.II. 2007. pp. 5-29

[11] Fatemi A, Glaum M. Risk management practices of German firms. Managerial Finance. 2000; **26**(3):1-17

[12] Wood A, Kellman A. Risk management practices by Barbadian banks. International Journal of Business and Social Research. 2013;**3**(5):22-23

[13] Hameeda AH, Jasim A-A. Risk management practices of conventional and Islamic banks in Bahrain. The Journal of Risk Finance. 2012;**13**(3):215-239

[14] Khalid S, Amjad S. Risk management practices in Islamic banks of Pakistan. The Journal of Risk Finance. 2012;**13**(2): 148-159

[15] Hassan A. Risk management practices of Islamic banks of Brunei Darussalam. The Journal of Risk Finance. 2009;**10**(1):23-37. DOI: 10.1108/15265940910924472

[16] Sifumba CM, Boitshoko Mothibi KB, Ezeonwuka A, Qeke S, Matsoso ML. The risk management practices in the manufacturing SMEs in Cape Town. Problems and Perspectives in Management. 2017;**15**(2): 386-403

[17] Miloš Sprčića D, Kožula A, Pecina E. State and perspectives of enterprise risk management system

development—The case of Croatian companies. Procedia Economics and Finance. 2015;**30**:768-779

[18] Miller KD. A framework for integrated risk management in international. Journal of International Business Studies. 1992;**23**(2):311-331 http://www.jstor.org/stable/154903 [Accessed 23rd June, 2016]

[19] Das TK, Teng B-S. Trust, control, and risk in strategic alliances: An integrated framework. Organisation Studies. 2001;**22**(2):251-283. DOI: 10.1177/0170840601222004 [Accessed 23rd June, 2018]

[20] Lowers M. 3 risk management practices of industry-leading organisations. In: The Risk Management Blog. 2015 http://blog.lowersrisk.com/risk-management-practices/ [Accessed 22nd June, 2018]

[21] Mitchell SL, Stern Switzer C. GRC Capability Model (OCEG Red Book). 2013. Lulu.com

[22] Stulz RM. Rethinking risk management. Journal of Applied Corporate Finance. 1996;**9**(3):8-25

[23] Marshall J, Heffes EM. Most firms agree: ERM is a challenge. Financial Executive. 2005;**21**(8):10

[24] Beasley MS, Branson BC, Hancock BV. COSO's 2010 Report on ERM. Committee of Sponsoring Organizations of the Treadway Commission; 2010

[25] Mitchell SL. GRC360: A framework to help organizations drive principled performance. International Journal of Disclosure and Governance. 2007;**4**(4):279-296

[26] Catmull E. How pixar fosters collective creativity. Harvard Business Review. 2009;**87**(1):109

[27] Kannan L, Thangavel H. Risk management lessons worth remembering from the credit crises of 2007. The Journal of Portfolio Management. 2008;**2008**:21-44

[28] Nocco BW, Stulz RM. Enterprise risk management: Theory and practice. Journal of Applied Corporate Finance. 2006;**18**(4):8-20

[29] Slywotzky AJ, Drzik J. Countering the biggest risk of all. Harvard Business Review. 2005;**83**(4):78-88

[30] Ehsan E. Risk management: The next source of competitive advantage. Foresight: The Journal of Futures Studies, Strategic Thinking and Policy. 2013;**15**(2):117-131. DOI: 10.1108/14636681311321121

[31] Porter ME. Competitive Advantage. New York, NY: The Free Press; 1985

Section 3

Risk, Technology and Engineering Contex

Chapter 6

The Internet of Things for Natural Risk Management (Inte.Ri.M.)

Riccardo Beltramo, Paolo Cantore, Enrica Vesce, Sergio Margarita and Paola De Bernardi

Abstract

This chapter deals with the development of a management system, which integrates the use of IoT in natural risk detection, revention, and management with economic evaluation of each stage. In the introductory part, recent data are presented that document the importance that natural disasters have for the environment and for the Italian economy. Section 2 presents the Inte.Ri.M. project—the Internet of Things for Natural Risk Management—its purpose, activity plan, and bodies involved. Technical aspects are treated in Section 3 with the choice of hardware and software components and the solutions for collecting and transmitting data. Section 4 is about the economic aspects considering the stages of prevention, intervention, and restoration and the relation between the intensity of human activity and environment to define a range of situations. These scenarios call for different economic methodologies useful to estimate economic implications of each stage in the short, medium, and long term. Section 5 describes the structure of the Inte.Ri.M. management system and the foreseen functionalities. In the conclusion, the critical points are discussed, and the steps for the transposition of the work carried out on the territory are outlined, according to the provisions of the work program.

Keywords: risk management, disaster, economic impact, sensor, forecast, IoT

1. Introduction

The structure of the chapter is described in **Figure 1**.

Evaluation and management of natural risks are relevant topics that have been brought to the fore by the recent earthquakes in Italy. The consequences of these phenomena make clear that it is mandatory to evaluate the risk and correctly manage information and emergency interventions. Securing the territory is on the top of the agenda of our government, and Piedmont region has recently adopted a plan to finance, with over 60 million euros, land protection and hydrogeological risk prevention measures [1].

Piedmont, a region in the north-west of Italy, located at the western edge of the Po Valley, is occupied for about 49% of its territory by the mountain ranges of the Alps and the Apennines, which delimit it on three sides like an arch. The region is statistically affected by alluvial events with average occurrences of an event every 18 months or so. In the alpine sector, special nivo-meteorological conditions can cause avalanches.

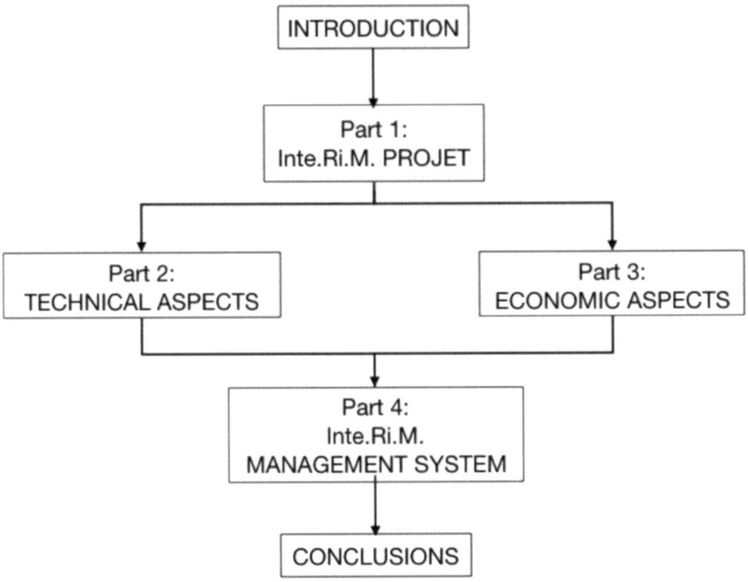

Figure 1.
The structure of the chapter.

The regional territory is also subject to earthquakes: the tectonic context and the active geodynamic regimes lead the region to be the site of seismic activity, generally modest in terms of energy, but notable as a frequency. Floods, fires, landslides, and avalanches can be less catastrophic compared to earthquakes, but they create damages as well, due to their frequency. Natural causes and an improper use of the territory are at the base of this kind of events. Management deficiencies cause **economic** losses in these contexts [2].

The effects of these phenomena influence human activities in the short and medium-long term; they cause real natural disasters whose damages are measured in victims and economic losses, in the short term, to the ecosystem and to human, civil, and productive settlements, in the medium-long term. The restoration of the initial conditions can take decades.

Although the attention devoted to forecasting and managing the effects of these phenomena has increased over time, and investments in parallel, the achievements recorded in the field of prevention have been few, and they show that the spaces for improvements exist and are wide. Monitoring networks are used where phenomena have occurred in the past or where there are significant threats of danger. As a result, large areas of territory are not monitored, while the availability of significant and updated data is the basis for any preventive action. The consequence of this attitude is that the interventions take place after natural disasters occur, while a preventive view, made up of infrastructural interventions, would allow to counter huge damages to the population and to the structures and to better organise the management of emergencies.

2. Inte.Ri.M. project

Based on this, the Inte.Ri.M. project—the Internet of Things for Natural Risk Management—has been launched in 2017, with an interdisciplinary approach, by a research group of economists, mathematicians, computer engineers, and experts in management systems of the University of Turin, linked to the Research Centre on

Natural Risks in Mountain and Hilly Environments of the same university and with the support of Piedmont region.

The choice of the acronym Inte.Ri.M. expresses the sense of provisional nature, which, in the specific case, can be associated to the dynamic balance of the elements of nature. Therefore, nothing is immutable, everything is temporary, and sometimes, the provisional nature is expressed in a brutal manner, with tragic consequences for humans.

The general objective of the project is the evaluation and management of natural disaster risks in a systemic and integrated way, in which technical and economic variables are measured to prevent and assess, with a holistic approach, the natural risks in ecosystem services.

The ultimate goal of the project is the realisation of a computerised integrated system for the evaluation of costs and benefits in some natural risk situations. The methodology that has been put in place is able to compare the costs of prevention, including the costs of the detection system, analysis and reporting, and an estimate cost to contain damage, with the benefits deriving precisely from the damage avoided (in the case study).

The specific objectives of the project, which correspond to the same number of work packages (WP), are as follows:

1. Identification of macro natural disaster risks in mountain ecosystem (WP1).

2. Consideration to disasters, whose impact is localised and circumscribed (WP2).

3. Focus on specific natural disaster risks: avalanches and landslides in Piedmont region (WP3).

4. Implementing technical tools and economic assessment approaches to prevent and manage natural disaster risks (WP4).

5. Application and integration of prevention and management system outputs on real-time dashboard with IoT tools to support disaster risk reduction decision-making processes (WP5).

6. Testing and applicability of the integrated key risk indicators selected from the dashboard in a pilot case (WP6).

7. Performance dissemination through communication and raise awareness of the institutions, businesses, and population interested in risk prevention (WP7).

Time sheet and deliverables are reported in **Figure 2**.

Currently, WP1 has been concluded and has led to the selection of sites to be monitored. WP2 and WP5 are in progress. Therefore, in the parts related to the technical aspects (Part 2) and to the economic aspects (Part 3), the analysis carried out to define the specifications of the integrated system is presented.

The identification of macro natural disaster risks in mountain ecosystem (WP1) has recorded an increment of natural disasters, in frequency and intensity. Evidence indicates that exposure of persons and assets to risk in all countries has increased and vulnerability has decreased, thus generating new risks and a steady rise in disaster-related losses, with a significant economic, social, health, cultural, and environmental impact in the short, medium, and long term. In this scenario,

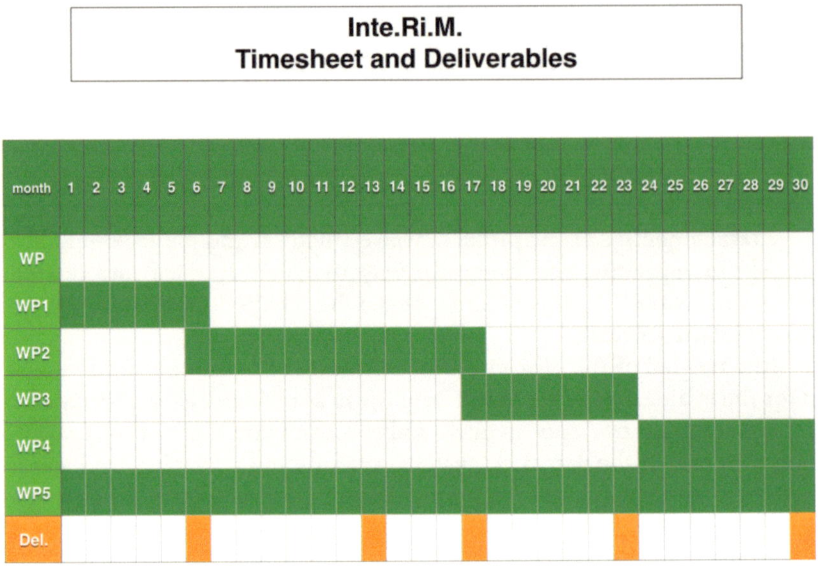

Figure 2.
Timesheet and deliverables.

mountain ecosystems, which furnish a large group of goods and services to humanity, for people that live in mountains or outside [3], like source of ecological and food security, are deeply affected by natural disaster [4].

It is widely accepted that mountain ecosystems must be protected but is neglected by the fact that they act also as buffer against natural hazardous. The measurement of full costs of losses and degradation of these ecosystem services are difficult to assess, but there is a sense of urgency to implement an economic tracking system to monitor all disaster risk reduction costs for mitigation, preparedness, and emergency response.

A correct evaluation of ecosystem services gives helpful information to estimate costs and benefits of decisions, to define future scenarios, and to recognise and avoid unexpected consequences [5].

The value of ecosystem services can be linked to the direct use (e.g., wood) or indirect use (value of specific ecological functions), or to the particular item relevance for other goods or values, or finally, the value may simply be intrinsic (e.g., existence value or cultural value). The capacity to prevent and respond to a disaster is especially important in mountains area, where remoteness and difficulty of access are often features of communities that, during disasters, are cut off from the outside world more often and for a longer time than lowland areas. For this reason, it emerges urgently the need to manage disaster risk with an effective integration among technical tools and economic approaches to measure the activities of prevention (ex ante risk management), rehabilitation, and recovery (ex post risk management) [6].

Environmental data are at the base of natural disaster prevention. The state of the art for landslide monitoring uses inclinometers, extensometers, and piezometer sensors. For avalanche monitoring, the most common solutions are ultrasound sensors, Doppler radars, and optic sensors. Floods and earthquakes require not just sensors; indeed, the most important prevision tools are software using mathematical algorithms based on time series. We will focus on avalanches and landslides in Piedmont region. The regional agencies of environmental protection of Piedmont (Arpa Piemonte) provide open data about river and snow levels, landslide areas, and other meteorological information.

Inte.Ri.M uses the data already available, integrating them with those measured by the WSN Scatol8®. Scatol8® is a WSN developed in the University of Turin.

Many sensors are included in the platform, that is, wind direction and speed, snow level, liquid flow, rain level, air pressure, solar and ultraviolet radiation, noise, vibration, soil moisture, and many others. Scatol8® platform is based on open source hardware and software not only for cost saving but also in the view of knowledge sharing. Several experiences have been made in environmental monitoring. We built WSN for data acquisition in mountain huts, parks, mine, and other kinds of areas.

Our interdisciplinary proposal aims to mix our technical and economic skills to assess how much preventing disasters with WSN can cost and make to save money at the same time. The use of open data that are already available is useful in the starting stage of the study and will reduce costs. At the same time, open source WSNs are suitable for detecting data, saving cost in comparison with proprietary solutions.

We consider economic and environmental indicators at the same time. Starting from the literature review, we choose the environmental variable relevant for the project scopes. We select the most suitable evaluation methods for the economic analysis of the ecosystem services (direct and indirect) that we integrate with the economic valuation of the human activities managed in the observed areas. The intensity of a measured variable or the result of statistical elaborations can assume different sense in the phase of sensing, data analysis, and communication with the stakeholders. Smartphone, tablet, and other devices are useful for the warning and the management steps of emergencies. They can also collect data from people to the emergency area. Assessing the economic impacts of disasters is a very recent field of study. The methodologies adopted to enable a transparent and coherent economic assessment of disaster risk management are based, first of all, on the post-event losses. The impact assessments estimate the economic and social impacts of past disasters. Storing such information in a database is a precondition for estimating future disaster impact. Moreover, we use the cost-benefit analysis (CBA) integrated with complementary tools as, for example, cost-effectiveness and multicriteria analysis (MCA) and robust decision-making approaches (RDMA). This holistic methodology is effective in relation to reliable input data availability (e.g., past disaster losses, indirect losses, and macroeconomic impacts), among which those from private business sector and public finance are important for achieving good-quality economic impact analysis. Approaches need to address the different layers of risk (from intensive to extensive risk), underlying risk drivers, as well as be tailored to mountain local contexts. Addressing these underlying risk drivers will reduce disaster risk, lessen the impacts of climate change, and consequently, maintain the sustainability of development (UNISDR, 2015). In parallel, attention will be paid to the need of communication among various actors involved in natural disaster management, through new technologies like the IoT and cloud computing.

3. Technical aspects

3.1 How technologies can contribute to disaster prevention and critical events notification

Disasters are caused by nature and/or by human activities, and they can impact on security, agriculture, industry, health, and natural environment. Technologies can have a key role in disaster prevention and management. While some tools, as for example, some environmental sensors, have been on the market since the first half of the twentieth century; in the last 10 years, the opportunity to connect devices through the Internet has created many new opportunities. People can use recent technologies for real-time analytics, remote or in-site monitoring, data analysis, early warning notification, victim localisation, and data aggregation. Research institutes are the organisations that commonly collect data and try to build some models useful for events forecasting. Those institutes can build their own data acquisition systems, and they can use data collected by others, for example, weather forecast by military or a mix of both. The diffusion of open data databases that people can consult for free on the Internet can help to have more data, but, at the same time, it is important to investigate about how those data were collected and how accurate are them.

Useful raw data can be collected not just from sensor networks but also thanks to IoT devices [7]. If we search information about disaster prevention and IoT, we can find that there is a lot of confusion in the use of the word IoT; unfortunately, this is true also in scientific papers. The biggest confusion is between Internet of Things and sensor networks. There are many definitions of IoT, and we refer to the one made by the Institute of Electric and Electronic Engineers. The document "Towards a definition of the Internet of Things (IoT)" [8] revision 1 (2015) tries to define the concept in 86 pages. We can find that the main concepts, taken from Section 5.3, are the following:

- **Interconnection of Things:** The first feature of IoT is derived from the name that describes it. It is a system that deals with the interconnection of "Things". The word "Thing" refers to any physical object that is relevant from a user or application perspective.

- **Connection of Things to the Internet:** From the name IoT, we can also learn that the "Things" are connected to the Internet. Accordingly, from the name we can deduce that the system is not an Intranet or Extranet of Things.

- **Sensing/actuation capability:** There is the involvement of sensors/actuators in the IoT system. The sensors/actuators are connected to the "Things" and perform the sensing and actuation, which bring the smartness of the "Things".

- **Embedded intelligence:** Smart and dynamic objects, with emergent behaviour, embed intelligence and knowledge functions as tools and become an (external) extension to the human body and mind.

- **Interoperable communication capability:** The IoT system has a communication capability based on standard and interoperable communication protocols.

The first point is important: the physical object itself must be relevant from a user or in the application perspective. For example, a temperature sensor that sends data to a remote dashboard is not an IoT device. On the other hand, a fridge that sense that the milk is ended and it informs as with a push notification or an alarm clock that changes the alarm time according to the traffic information are examples of IoT devices.

Figure 3 shows how data can be collected and used. Data sources can be sensors, other databases, or people notifications. In the last case, people can contribute in many ways, for example, using a mobile application to notify something.

Once raw data are acquired, the business logic can notify alarms or create new data, for example, forecasts, combining trends, and different data sources. The intensity of a measured variable or the result of some statistical elaborations can assume different sense in the phase of sensing, data analysis, and communication to the stakeholders.

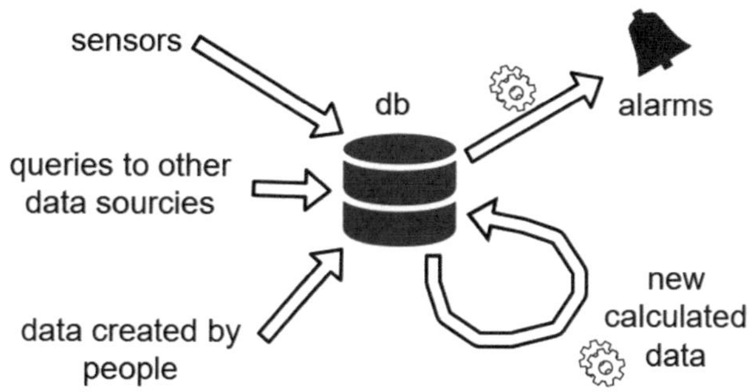

Figure 3.
Data input and output.

3.2 Environmental risk and sensors and forecasting technique examples

Environmental data are at the base of natural disaster prevention. In this section, we will explore what are some of the best sensor-based solutions currently on the market that allow to predict or to early warning disasters.

3.2.1 Landslides

The state of the art for landslide monitoring uses inclinometers, extensometers, and acceleration and piezometer sensors. Before using sensors, it is important to have an inventory with the zones of interest. This can be done to detect landslide that already happened with high-resolution aerial images [9] or LiDAR data. These data must be aggregated to the information that researchers can obtain from city registers, people interviews, other researcher's databases, etc. Once the areas of interest are defined, it is possible to use sensors for advance warning, for event warning, or for both the scopes together.

Inclinometer (see **Figure 4**) and extensometer can be used for advance warning, detecting the small changes in the distance between two points. The most common idea [10–13] for landslide monitoring during the event is to put a lot of nodes on the area of interest; in this way, the signal can be forwarded to the neighbour nodes before the sending nodes probably die [10-13].

3.2.2 Avalanches

Compared to other natural disasters, the risk of avalanches is more to be controlled by humans. Many studies correlate the temperature changes, the type of the snow, and the probability that an avalanche occurs. Humans can also create new controlled avalanches to avoid a disaster, for example, they can trigger some small avalanches before too much snow accumulates on the mountains over a part of a ski resort or a street. In any case, human's understanding of the dynamic response of the snowpack to avalanche control explosives is rather limited [14].

The historical and present weather conditions can be used to forecast the snow conditions in the future. In Europe, the model COSMO (http://www.cosmo-model.org/) is an example of a project developed to create an atmospheric forecasting model. Ref. [15] uses this model in conjunction with a snow cover model to forecast avalanche activity; particularly, they concentrate on wet-snow instability. The WSL Institute for Snow and Avalanche Research SLF creates a repository that includes the snow cover model that they used and other similar ones (https://models.slf.ch/).

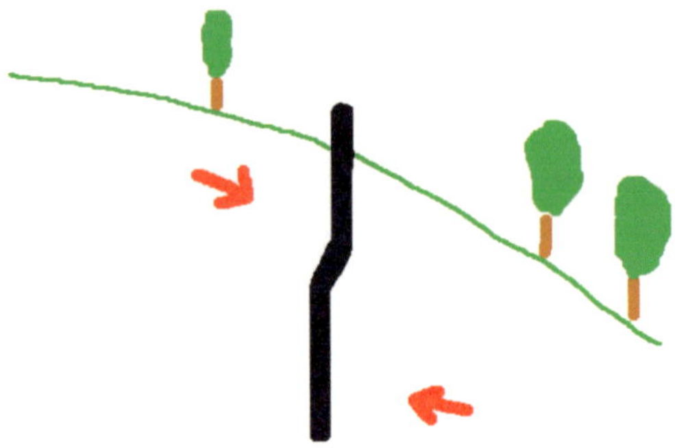

Figure 4.
Inclinometer probes allow to detect ground movements under the soil.

If the historic trend of environmental parameter such as temperature and snow level is really important to discover the avalanche's probability, in situ test made by people can be also quite important to discover how much the layers of snow are discontinuous.

For avalanche monitoring, the most common sensor solutions are **ultrasound sensors**, **Doppler radars**, and **optic sensors** [16]. Other solutions include **cameras** and image recognition, LiDAR, or the use of **geophone** built for earthquake monitoring [17]. These seismic sensors have two roles: monitoring the activity of the avalanche and studying the dynamics of the avalanche. It is interesting to notice that every avalanche does not only produce a seismic signal but also some low frequency infrasonic acoustic waves (below 20 Hz). Remote infrasound solutions can be used for monitoring also when the visibility is low [18].

Using geophones, it is important to discriminate signals due to the avalanche from other noises. The analysis in the time-frequency domain allows to distinguish different sources, for example, helicopters or airplanes that generate signals on some specific harmonic frequencies [19]. It is also interesting to notice that diverse kinds of avalanche; for example, loose snow and slab ones have their typical signal characteristics. Detecting precursor signal with geophones is very difficult but these instruments allow not just to detect avalanches but also to study the dynamic of the snow inside the avalanche.

Another interesting aspect of snow to be monitored is the snow flow. The snow can be moved by the wind, and in this way, it can influence the avalanche danger, it can block some roads or the access to some buildings, and it can cause other problems. Lehning and others, in the article *Snow drift: acoustic sensors for avalanche warning and research*, compare many kinds of sensor for snow flow; in particular, they concentrate their analysis on the acoustics sensors. Their findings are that, at the present, it is not possible to measure the mass of the flux of snow with high accuracy [20]. A significant reduction of false detection in infrasonic measure can be obtained using machine learning classification techniques compared to the use of a threshold [16].

One problem of snow monitoring is also that, in all the previous cases, you can only monitor a specific area, but the snow condition can be quite different in another place not far from the installed sensors.

New technologies have an important role not just in avalanche detection but also for localising the victims. Common instruments like the ARTVA allow to find people under the avalanche. New tools like wearable sensors are currently not so diffused, but in future, they will probably help the rescue operations, giving information about the health status of the victims under the snow [21].

3.2.3 Floods

The risk of flood disaster increased in the last years because people build more buildings on floodplains, and the area covered by cement has increased. To make actions against flood effective, it is important to forecast it further in advance as possible. Usually people working in the emergency response staff are notified before common people. This happen because early previsions are more subject to false alarm [22]. The predictability can change a lot depending on the historical data and the place. For example, Webster et al. in their article demonstrate that it was possible to predict the 2010 floods in Pakistan 6–8 days in advance using a multi-year analysis and a hydrological model to mitigate the impact anticipating actions [23]. Rain sensors alone cannot forecast floods; the most important prevision tools are software using mathematical algorithms based upon time series. The data for the analysis can be a mix among rain gauge measures and satellite information. Prediction models useful for short-range prediction can differ from the ones that are best for long-term forecasts [24]. It is interesting to notice that simulate flood disaster in cities must also take into account the sewer capacity [25].

Other types of events to be forecasted for safety reasons are the wave run-up and the wave overtopping. The project HIDRALERTA [26] developed for the safeguard of the Portugal coasts is an interesting case because it is a tool for evaluating the risk, taking a decision, and eventually alerting the involved stakeholders. The explained methodology is composed by four parts and can be applied to different geographic area with different data sources:

- The wave characterisation: This step can be accomplished with some spectral or mid slope wave models. These predictive models need the sea to be monitored and, being too much complex, can be simplified using some matrix with some pre-run different models for multiple conditions. The use of machine learning techniques for predicting values can also help.

- A neural network determines if the wave will overtop the barrier or not.

- Risk level calculation for the overtopping values. Risk maps are produced taking into account the thresholds for the overlapping, the probability for the event to occur, and the consequences.

- The warning and alarm system.

3.2.4 Earthquakes

As for floods, also for earthquakes, the most important prevision tools are software using mathematical algorithms based upon time series. Earthquakes can have a bigger impact if they affect some human structures like a dike or a nuclear power plant. Herle and others [27] simulate many dangerous events for calculating the risk of a dike failure. In their simulation, they use a protocol called message queue telemetry transport (MQTT) that is a lightweight IoT transmission protocol. We can find an interesting approach in the article by Zambrano et al. [28]. They build a system for early detecting and notifying earthquakes using people smartphones' accelerometers. The work is a notable example of a project that can be developed almost without having new costs and using existing devices for a new scope. Munib ur Rehman and others in their article analyse how ICT and wireless sensor networks can contribute to detect earthquakes early. They used three main different approaches correlating three events with the possibility of an earthquake. Detecting abnormal animals' movement can help to early prevent calamity, and also variation in groundwater pressure gives information about the possibility of earthquakes. Another precursor is the concentration of radon on the ground that changes because of the breakings of rocks below the surface.

Sensors can give an important contribution also after a disaster; for example, they can monitor pollutants and contaminants including radioactive scenarios after an earthquake has damaged an industrial area.

3.2.5 Debris flows

Another dangerous process that can occur during strong rainfalls is the debris flow. Compared to landslides, the debris flows are more difficult to study because these events can occur in a very short time, and using aerial photographs, it is hard to find traces of previous events [29]. An approach for forecasting debris flows can be the creation of a map with the alluvial fan using LiDAR data and photograph and keeping trace of the event occurrences. Because this event can be considered as the union of landslides and water floods, sensors can help to detect strong waterfall and notify early warning. Example of sensors for this application is ultrasonic or radar gauges, ground vibration sensors, video cameras, avalanche pendulums, photocells, and trip wires [30].

3.2.6 Fire

It is very hard to forecast a fire. It can be caused directly by humans or by some objects like trains or electrical lines. If we want to limit fire consequences, it is important to detect the fire as soon as possible. A possible approach is to use IP cameras to detect the smoke at its beginning using some detection algorithms [31, 32, 33]. Another approach is to use geostationary satellites and image processing [34].

Fire can be early detected even before the smoke presence using gas sensor as demonstrated by Gutmacher et al. [35] in the article "Gas Sensor Technologies for Fire Detection". In their experiment, the fire was detected by gas sensors 4 minutes before one optic smoke sensor detected it. This approach can be used mostly in closed areas, for example, offices and houses. In the same context, also heat detectors can provide early earning of fire incidents.

An interesting experiment in outdoor context [36] uses the wireless sensor network transmission signal strength indicator variance between two nodes to detect the fire. The idea is that the changes in the vegetation water content level due to the fire influence the wireless signal.

3.3 Other scenarios and solutions

Human activities can have an appreciable impact on natural disaster activities can have as a consequence relevant natural disasters, for example, the Chernobyl nuclear disaster that happened during some scheduled test. Also, in the Seveso disaster, people working in the company had many responsibilities [37]. Apart from human dangerous behaviours and errors, when an organisation is conducting an environmentally risky activity, it is important to continuously monitor the relevant processes.

Some very diffused technologies can help during disasters: smartphone, tablet, and other devices are useful for the alert and the management steps of emergencies. They can also collect data from people in the emergency area, for example, the device can share the position of an injured person using the GPS. Also, social network like Twitter and Facebook can help to collect data from people in the zone of the emergency.

Lapante and others in the article "Could the Internet of Things Be Used to Enhance Student Nurses' Experiences in a Disaster Simulation?" [38] illustrate the simulation of a street accident with many people injured. Nurses and patients in the place are connected and can share information about the patients, thanks to some wearable sensors and the use of RFID for patient recognition. If their approach is not realistic because people are not usually equipped with a RFID system for recognition, an interesting approach is the one of Libelium Meshlium. This product, which is currently on the market, allows to count people in a space looking for the Bluetooth and Wi-Fi signal of their mobile phones. This approach cannot assure to count the total amount of persons, but it is easy, and it is applicable in real word and not just in simulations.

3.4 Possible issues in data communication

Many technologies are useful for collecting data from the sensor nodes in a network. **Table 1** shows some of the currently used communication technologies. In the most common scenario, the Internet network is used to transmit data to a remote server. It is important to underline that different protocols can operate together: for example, we can have one protocol for data collection among the nodes of a wireless sensor network (WSN) and another one from one collector node in the WSN and the Internet.

The data communication is easier if we collect data for doing forecast but is more critical if we need to transmit information in real time during the disaster. The main problems in transmitting data in such conditions are the availability of power supply for the nodes and of the data network for transmitting the data packets.

Many research reports (e.g., [39]) discuss the importance of redundancy. It is possible to use some battery to assure that the node will transmit data also if there is no external power, but the most critical point happens if the data network is absent. For example, if the cellular base transceiver station of our operator is not available, it becomes important to use the one of another operator. All the mobile phones permit to make emergency call independently from the operator and the credit, but this is not the same for other services like sharing the position.

In some applications, for example, avalanche monitoring, we can put many radios to have more chances that almost one radio will successfully transmit. The adoption of a mesh network, where a packet can take many paths to reach its destination, also helps.

The importance of having a working system during the disaster depends on the specific application. For example, at that time, information needed for prevention can be less useful with respect to data useful to localise injured people overwhelmed during an earthquake. The network services can be unavailable not just because the radios are physically broken but also because there are too many requests: for example, many people trying to call with their mobile phones just after the disaster can saturate the cell.

Before proceeding with the installation of the monitoring network prototype, it is considered useful to test the performance of the network in a mountain environment. For this reason, the site of the Angelo Mosso Institute of the University of Turin has been chosen. It hosts groups of international researchers working in various scientific areas. The objective of this phase is the verification of the energy performance of the remote sensing network in relation to the frequency of data transmission [40].

Name	Network topology	Average ranges	Average Tx power	Scenario
ZigBee (s2/s2pro)	Mesh	20 m/3.2 km	3.1 mW/63 mW	Many nodes, node routing needed
3G/4G	ptp (from users view)	Typical carrier ranges 30 km	2 W	Direct communication to a remote server thought the Internet. No LAN available
xBee 802.15.4 (s1/ s1 pro)	Fully connected star	120 m/3.2 km	10 mW	Low power, not many nodes
xBee s5 868/900	Star	500 m–12 km	315 mW	Have to collect data in open spaces. Do not need packet routing
Lora	Typically star (mesh supported)	5–15 km	100 mW	Need to transmit small amount of data covering distances in Km order. Need a WAN gateway in the same technology as the nodes
Sigfox	Star	30 km	300 mW	Direct communication to a remote server thought Internet, less expensive than 3G. Only few access points available
Satellite	ptp	36,000 km	2–5 W	There are no other ways to connect. For example, you are in the sea or in a desert
Wi-Fi	Typically star	50 m	100 mW	Can use an already built Wi-Fi network. Do not need packet routing
Bluetooth	star	50 m	10 mW	Need to interface with a Bluetooth device

Table 1.
Technologies suitable for data transmission in disaster scenario.

4. Economic aspects

The economic part of the Inte.Ri.M project aims at selecting correct methodologies to evaluate and to allow the comparison among costs of prevention and the damages from the disasters that can be avoided. In order to achieve this ambitious goal, an overall strategy is defined, and three parts are identified. The first is about the different scenario to analyse, depending of the type of territory that disaster may affect, and the second is related to the different methodologies used for evaluation. The third part is focused on the actual possibility to compare the costs and benefits of disaster prevention.

4.1 Territory and disasters

According to the area that disasters affect, it is possible to identify different situations. Inte.Ri.M. project has focused its attention on mountain ecosystem (Piedmont region) and above all on specific natural disaster: avalanches and landslides, with circumscribed and localised impact. As seen, the same reasoning can be extended to other geophysical (like earthquakes and forest fires) and hydrological disasters [41].

For the purposes of ongoing research, two classification parameters have been identified: the *type of territory* affected by the disaster and the time taken to recover from the consequences. Both are important for economic evaluations.

The first, it is fundamental to choose the correct variables to be considered. It is very different to destroy an urban area from a completely uninhabited area. Thus, as a sample area, two different types were chosen, always in a mountain environment: a highly natural area and an urbanised area.

In **Figure 5**, the presence of the ecosystem services (ESS) or economic components can be noticed, depending on the type of area taken into consideration.

The second, it is useful to evaluate the "time axis" that sets the economic reasoning. **Table 2** shows the various recovery times for different disaster and for type of area.

This subdivision based on the time horizons allows, within the economic assessment, to make a more correct comparison with the benefits of the avoided damage. To consider that the comparisons must be homogeneous also from the temporal point of view with the benefits that derive from the systems that disasters are going to disturb. In the case of ecosystems, as well as in traditional economic systems, the perspective is that of the medium/long term (Ministero dell'ambiente e della tutela del territorio e del mare, 2009).

First of all, we identified what are the information concerning natural disasters to be taken into consideration. After that, we proceed to build a scheme useful for economic evaluation.

To achieve this, it is necessary to hypothesise which valuation objects would have been damaged and, on the basis of this, verify which tools are available to give an economic value. Ecosystems played an important role in this process. Even if they can be classified into natural, semi-natural, and artificial [42], for the natural territories, the presence of natural ecosystems was particularly assumed, and therefore, we have taken into consideration the damage only to this system. Therefore, the urbanised areas are not considered as artificial ecosystems (urban ecosystem), and other methodologies are used to calculate damages on people, infrastructure, and industrial plants.

4.2 Disaster on highly natural area: the value of ecosystem service (ESS)

When it was agreed that economy and ecology had to go in tandem (cfr "ecological economy") against a background of unsustainable growth in the resources use, common language has been sought. The first step to solve this challenge is to refer to a common functional unit: the ecosystem (Ministero dell'ambiente e della tutela del

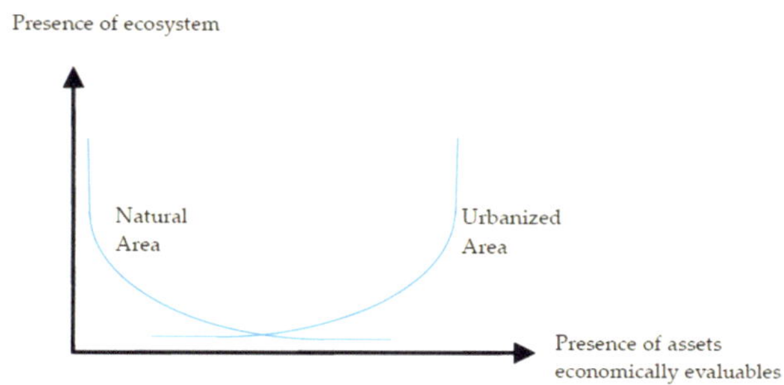

Figure 5.
Presence of ecosystem or assets economically evaluables.

	Short term	Medium term	Long term
Natural area	Earthquakes, avalanches	Landslides	Landslides
Urbanised area		Landslides, avalanches	Earthquakes, landslides, avalanches

Table 2.
Time axis for disasters and areas of different types.

territorio e del mare, 2009). Because it is very difficult to evaluate an ecosystem for the multiple relationships inside, we value ESS to human well-being [43].

This aspect gives us a chance on developing a different logical argument. In fact, rarely, the discussion on an area's value is focused on ESS: normally all attention is on the role that plays on local economies. Areas have to be recognised for the service they provide: mitigation of climate change and natural disasters, disease control, maintenance of water quality, and cultural services, including recreation, maintenance of historical or iconic landscapes, and protection of sacred natural sites [44].

ESSs are defined by the Millennium Ecosystem Assessment as "the benefits people obtain from ecosystems" [3]. Fisher and Turner [45] added that "ecosystems services are the aspects of ecosystems utilised (actively or passively) to produce human well-being". The MA developed and claimed a very important concept: the well-being of society is strictly linked to services provided by nature. **Figure 6**, by MA [3], shows this relation.

The measurement of losses' full costs and degradation of these ESSs are difficult to assess, but there is a sense of urgency to implement an economic tracking system to monitor all disaster risk reduction costs for mitigation, preparedness, and emergency response. A correct evaluation of ESS gives helpful information to estimate costs and benefits of decisions, to define future scenarios and to recognise and avoid unexpected consequences [5]. The value of ESS can be linked to the direct use (e.g., wood) or indirect use (value of specific ecological functions) or to the particular item relevance for other goods or values, or, finally, the value may simply be intrinsic (e.g., existence value or cultural value). To narrow the field and make a good comparison, the first step is to draw from the studies that have circumscribed key ESS of the different biomes on Earth [3]. The starting point is the ecosystems present in the investigated areas. The logical path is that of analysing the identification of the services offered, to move on to quantification and valorisation [46]. In this case, we consider ESS in alpine environments (mountains and forests), which mainly serve areas with high naturalness but also urbanised areas, always in a mountain environment. For the *identification* step, it is important to start from ESS definition. If ESSs are useful for human well-being, the supply of and demand for ESS have to be determined [47]. In order to define which type of ESS the territory can supply, the reviews cited by Haida et al. [48] have reported that most investigated are regulating services (mainly carbon sequestration) followed by provisioning services (such as the supply of fresh water and food) and cultural services (largely recreation and tourism). From the demand side, an analysis of inhabitants, touristic flows, and activities must be conducted (on bibliographic basis). The steps of *quantification* and *evaluation* show the main difficulties to arrive at a concrete and usable result. Starting from Costanza's study, in one of the most famous ecological economists, where dozens of scientific papers on economic evaluation of ESS [43] are summarised [43] and produce a "summary of average global value of annual ecosystem services", many other researches have tried to get into the specifics of this complex assessment.

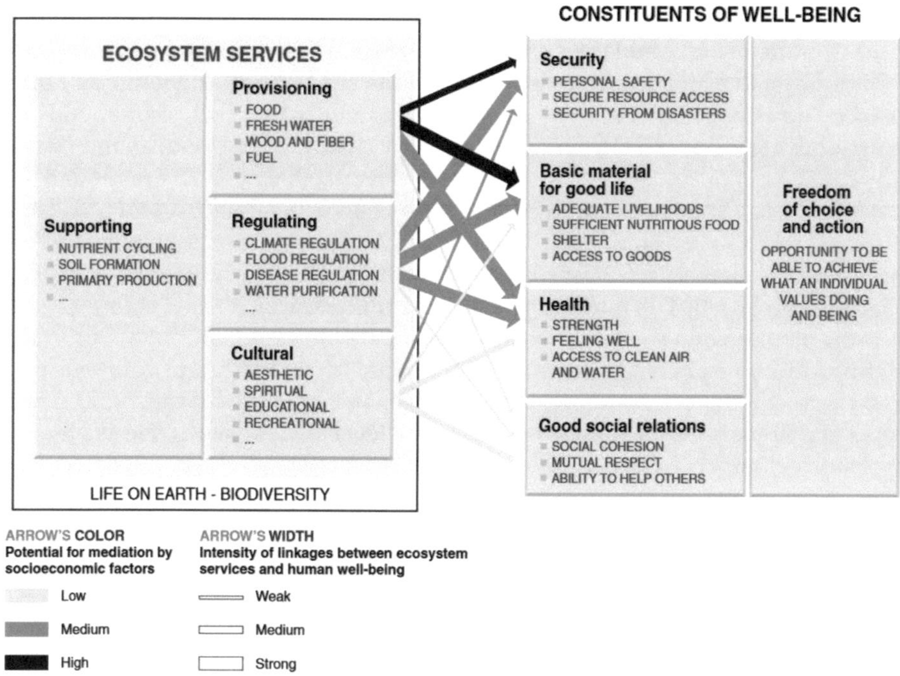

Figure 6.
Ecosystem services and human well-being.

Despite all the progresses made in this field, it is difficult to assess environmental assets, especially when it is necessary to attribute value to the benefits they generate. Even if the monetary evaluation of environmental goods without a market can be more or less correct, it provides a starting point for hypothesising policies and strategies. In order to focus attention and attribute a monetary value to environmental goods and, in this case, to ESS, the willingness to pay (WtP) and total economic value (VET) are often used. Also, in his review reports, WTP of individuals for ESS is the first method used [43]. In this case, it is emphasised that it is not a matter of attributing a value to a real exchange but only to make comparisons and establish strategies. Monetary value can be used as an indicator to encourage development policies.

Instead, as in the case of PES (payment for ecosystem services) [46], governments and private enterprises have started to pay for ESS [44]. Other authors have stated that its price does not represent the good real value but only one way to facilitate comparison with actions that have a market [49].

For quantification and attribution of value in the development of the project, reference was made to the data sheets deriving from Making Good Natura (LIFE+11 ENV/IT/000168) [5]: within it is possible to find indications for the evaluation of different ESSs.

4.3 Disaster on urbanised area: the value of life, infrastructures, and plant

In this part, the urbanised area is taken into consideration, but "urbanised like" can be in mountain territory, the focus of the Inte.RI.M. project. The urban situation is easier to assess but only apparently. It is true that there are methods for evaluating human lives and economic activities, but for these too, there are aspects of uncertainty as well as those predicted for ESS.

In order to monetise the value of a *human life* in disasters, we often refer to VOSL (value of statistic life) and DALY (disability-adjusted life year) [50].

For a "complete and detailed assessment of the exposure of 'economic activities' to natural disaster", Marin et al. study [51] refers to literature that recommends the use of different proxies depending on type of disaster. For flood, an interesting study carried out on an Italian case exploits the overlap between risk maps and residential areas. Starting from this information and using databases on real-estate values and reconstruction costs, it is possible to arrive to a damage assessment "ex ante" in case of floods useful for strategic reasoning [52]. For landslides, another Italian case is chosen. The interesting aspect of this research is that "Four types of assets are combined in the weighted assets map: (1) physical (buildings, cultural buildings and transportation networks), (2) social (population), (3) economic (land value), and (4) environmental (land cover and protected areas)" [53]. The joint evaluation of these variables on a concrete level allows us to arrive at a hypothesis that is not too far from reality.

4.4 Balanced scorecard

For monitoring and managing economic aspects of the project, we use different methods and tools that belong to the manager's toolbox, for example, cost-benefit analysis and multivariate analysis (in this work, we will focus on the former). But rather than using these raw mathematical, statistical, or financial tools alone, we integrate them in a wider framework, balanced scorecard, whose strengths are very useful for the Inte.Ri.M. project.

Balanced scorecard is a business framework used for tracking, planning, and managing an organisation's strategy. The knowledge of balanced scorecard was introduced by Robert S. Kaplan and David P. Norton in 1992 [54] and spread thanks to their book in 1996 [55]. The many criticisms the framework received are not discussed in this work.

Many benefits come from balanced scorecard. Among all, the framework allows:

- Transforming strategic objectives in operational objectives
- Defining priorities of development projects and actions
- Measuring and checking organisation's path towards defined targets
- Improving competencies in building and implementing strategies
- Spreading knowledge of the strategy in the whole organisation

Balanced scorecard is based on four "perspectives": financial, customer, internal business processes, learning, and growth. For each one, relevant topics are objectives, measures, targets, and initiatives, which must be defined, described, or computed as a preliminary step in strategy definition.

Mainly used in business, balanced scorecard is also used in the public sector, that is, government and non-profit organisations. In these cases, perspectives may be reordered differently or adapted to this different context. Specific benefits for government or non-profit organisations arise from strategy maps, a sort of add-on for balanced scorecard [56]. Strategy maps emphasise cause and effect relationships among perspectives and linked objectives but specifically take into account human capital and other variables belonging to the field of knowledge economy that usually financial accounting does not consider. This is the most important reason why we decide to use the balanced scorecard with strategy maps. In the Inte.Ri.M. project, each type of area has a distinct balanced scorecard.

4.5 Cost-benefit analysis and cost-effectiveness analysis

It is in this framework that we integrate classical management tools like cost-benefit analysis [57, 58], for comparing prevention costs with the damages from the disasters that can be avoided. One of the preliminary activities required to perform cost-benefit analysis is to measure all cost and benefit elements and to convert them in some unit of monetary measurement. Then, traditional indicators like net present value or internal return rate are computed.

In many applications of the analysis, especially in the public sector, problems arise because elements are present that analyst or policymaker cannot or does not want to convert in a monetary measurement, that is, human lives, healthcare quality, or beauty of a landscape. For an example about environmental resources, close to our research interest, see [59]. For this reason, besides cost-benefit analysis, we also use cost-effectiveness analysis. Although typically employed in health applications, the Inte.Ri.M. project can benefit from this technique as it allows to compare alternatives based on their cost and a measure of effectiveness that is quantified but not monetised, for example, the number of lives saved instead of a monetary value of lives. Given the impossibility to compare costs expressed in monetary units and benefits expressed in another unit of measurement, cost-effectiveness analysis proceeds through the construction of cost-effectiveness indexes that allow to compare the various alternatives.

Although the researcher's toolbox is very rich, some problems still need to be investigated. One of the most critical variables in the Inte.Ri.M. project is recovery time. Recovery time is neither a cost nor a benefit element, but it is essential to define the time window costs, and benefits are spanned over. Recovery time is difficult to estimate and is likely to increase. To make matters worse, an increase in recovery time induces changes in amount and allocation in time of monetary values. This dependence on an exogenous variable requires to reconsider the financial model and may impact considerably on results and on sensitivity analysis, one of the final steps of cost-benefit analysis.

5. Conclusions

The chapter outlined the state of the art of Inte.Ri.M. project. The on-site activity will provide answers to the question marks determined by the specific operational conditions established by the mountain environment.

The project will be able to generate benefits that potentially will have an impact in terms of creating value for the area. Beneficial effects are identified on two levels:

- A broad level: regarding the method of analysis development that integrates economic and technical evaluations. This output is designed to be made replicable in as many areas as possible.

- The restricted level (identified as "pilot case") that through the project will make use of hard data for risk management in the phases of prevention, intervention, and recovery is capable of enhancing the data relating to the analysis of costs and benefits.

The project development is planned at two levels (national and international). Each stage foresees a growing extension (qualitative and quantitative) of Inte.Ri.M. system through applications that imply an increasing complexity, in terms of technical, economic, and management issues to cope with and in number and variety of partners to involve. The needs of time and funding grow, therefore, consistently. The output of our project and actions planned to increase interest and promote cooperation, jointly with the research team's propensity to act internationally, could make international level directly accessible.

Author details

Riccardo Beltramo*, Paolo Cantore, Enrica Vesce, Sergio Margarita and Paola De Bernardi
University of Turin, Turin, Italy

*Address all correspondence to: riccardo.beltramo@unito.it

References

[1] Piemonte Region, Piemonte Informa [Online]. 2017. Available from: http://www.regione.piemonte.it/pinforma/ambiente/1695-oltre-60-milioni-per-la-difesa-del-suolo-e-contro-il-dissesto-idrogeologico.html. [Accessed: 20-7-2018]

[2] Arpa, Piemonte Region, Rischi Naturali [Online]. 2018. Available from: http://relazione.ambiente.piemonte.it/2018/it/territorio/fattori/rischi-naturali. [Accessed: 20-7-2018]

[3] Millenium Ecosystem Assessment. Ecosystem and Human Well-Being. Washington, DC: Island Press; 2005

[4] Nibanupudi HK, Shaw R. Mountain Hazards and Disaster Risk Reduction. Japan: Springer; 2014

[5] Schirpke U, Scolozzi R, De Marco C. Modello dimostrativo di valutazione qualitativa e quantitativa dei servizi ecosistemici nei siti pilota. Parte1: Metodi di valutazione, Bolzano. 2014;**1**:3-5

[6] Zimmermann M, Keiler M. International frameworks for disaster risk reduction: Useful guidance for sustainable mountain development? Mountain Research and Development. 2015;**35**(2):195-202

[7] Gaire R, Sriharsha C, Puthal D, Wijaya H, Kim J, Keshari P, et al. Internet of things (IoT) and cloud computing enabled disaster management. In: Integration of Cyber-Physical Systems, Cloud, and Internet of Things. New York, United States: Cornell University Library; 2018

[8] IEEE. Towards a Definition of the Internet of Things (IoT) [Online]. 2015. Available from: https://iot.ieee.org/definition.html

[9] Yan Q, Li H, Jing L, Ge W, Ding H, Tang Y, Kongwen Z. Research on Automatic Detection of Seismic Landslides Information. IEEE International Geoscience and Remote Sensing Symposium (IGARSS). Beijing, China: IEEE; 10-15 July 2016

[10] Mishra PK, S. K. Shukla Dutta S, S. K. Chaulya Prasad GM. Detection of Landslide Using Wireless Sensor Networks. IEEE; 2011

[11] Supekar PR, Takayama S. Characteristics evaluation of sensing network in wireless landslide surveillance system. In: Proceedings of the SICE Annual Conference; Tsukuba, Japan. 2016

[12] Khoa VV, Kazuya O, Shigeru T. Collaboration of host system and local sensing node network system. In: Proceedings of the SICE Annual Conference; Tsukuba. 2016

[13] Vera JE, Mora SF, Cervantes RA. IEEE Biennial Congress of Argentina (ARGENCON). Design and testing of a network of sensors on land surfaces to prevent landslides. Buenos Aires Argentina; 2016

[14] Hans-Peter M, Schweizer J, Birkeland K. Some recent advances in snow and avalanche science. Cold Regions Science and Technology. 2011:119-121

[15] Bellairea S, Herwijnena AV, Mittererc C, Schweizera J. On forecasting wet-snow avalanche activity using simulated snow cover. Cold Regions Science and Technology;**2017**:2838

[16] Thüring T, Schoch M, Herwijnen AV, Schweizer J. Robust snow avalanche detection using supervised machine learning. Cold Regions Science and Technology. 2015:60-66

[17] van Herwijnen JSA. Monitoring avalanche activity using a seismic

sensor. Cold Regions Science and Technology. 2011:165-176

[18] Scott E, Hayward C, Kubichek R, Hamann J, Pierre J, Comey B. Single and Multiple Sensor Identification of Avalanchegenerated. Cold Regions Science and Technology. 2007 pp. 159-170

[19] Herwijnen AV, Turner J, Schweizer J. Listening to Snow—Avalanche Detection Using a Seismic Sensor Array. Squaw Valley, CA: International Snow Science Workshop; 2010

[20] Lehning M, Naaim F, Naaim M, Brabec B, Doorschot J, Durand Y, et al. Snow drift: Acoustic sensors for avalanche warning and research. Natural Hazards and Earth System Sciences. 2002:121-128

[21] Michahelles F, Matter P, Schmidt A, Schiele B. Applying Wearable Sensors to Avalanche Rescue: First Experiences with a Novel Avalanche Beacon. Computers and Graphics. Pergamon New York: 2003;**27**(6)

[22] Stephens E, Cloke H. Improving flood forecasts for better flood. The Geographical Journal. 2014:310-316

[23] Webster PJ, Toma VE, Kim H. Were the 2010 Pakistan floods predictable? Geophysical Research Letters. 2011:1-5

[24] Wu L, Seo D-J, Demargne J, Brown JD, Cong S, Schaake J. Generation of ensemble precipitation forecast from single-valued quantitative. Journal of Hydrology. 2011:281-298

[25] Lee C-S, Ho H-Y, Lee KT, Wang Y-C, Guo W-D, Chen DY-C, et al. Assessment of sewer flooding model based on ensemble quantitative precipitation forecast. Journal of Hydrology. 2013:101-113

[26] Sabino A, Poseiro P, Rodrigues A, Reis MT, Fortes CJ, Reis R, et al. Coastal risk forecast system. Journal of Geographical Systems. 2018;**20**:159-184

[27] Herle RBJBS. Smart sensor-based geospatial architecture for dike monitoring. In: IOP Conf. Series: Earth and Environmental Science; Saint Mary; 2016

[28] Zambrano AM, Perez I, Palau C, Esteve M. Distributed sensor system for earthquake early. IEEE Latin America Transactions. 2015;**13**:291-2998

[29] Margottini C, Canuti P, Sassa K, et al. Landslide Science and Practice: Volume 1: Landslide Inventory and Susceptibility and Hazard Zoning. Springer, Verlag Berlin Heidelberg; 2013

[30] Arattano M, Marchi L. L. Systems and sensors for debris-flow monitoring and warning. Sensors. 2008:2436-2452

[31] Millan-Garcia L, Sanchez-Perez G, Nakano M, Toscano-Medina K, Perez-Meana H, Rojas-Cardenas L. An early fire detection algorithm using IP cameras. Sensors. 2012:5670-5686

[32] Kanwal K, Liaquat A, Mughal M, Abbasi AR, Aamir M. Towards development of a low-cost early fire detection system using wireless sensor network and machine vision. Wireless Personal Communications. 2017:475-489

[33] Qureshi WS, Ekpanyapong M, Dailey MN, Rinsurongkawong S, Malenichev A, Krasotkina O. QuickBlaze: Early fire detection using a combined video processing approach. Fire Technology. 2016:1293-1317

[34] Koltunov A, Ustin SL, Quayle B, Schwind B, Ambrosia VG, Li W. The development and first validation of the GOES early fire detection (GOES-EFD) algorithm. Remote Sensing of Environment. 2016:436-453

[35] Gutmacher D, Hoefer U, Wöllenstein J. Gas sensor technologies

for fire detection. Sensors and Actuators B: Chemical. **2012**:40-45

[36] Qiang C. A forest early fire detection algorithm. Sensors & Transducers. 2014:73-79

[37] Centemeri L. Retour à Seveso: La complexité morale et politique du dommage à l'environnement. Annales. Histoire, Sciences Sociales. 2011:213-240

[38] Laplante NL, NL PA, Laplante PA, Voas JM. Could the internet of things be used to enhance student nurses' experiences in a disaster simulation? Online Journal of Nursing Informatics. 2018

[39] Islam RU, Andersson K, Hossain MS. Heterogeneous wireless sensor networks using CoAP and SMS to predict natural disasters. In: 2017 IEEE Conference on Computer Communications Workshops (INFOCOM WKSHPS); Atlanta, GA; 2017

[40] Istituto Angelo Mosso. Università degli studi di Torino [Online]. Available from: https://www.unito.it/ateneo/strutture-e-sedi/sedi/piano-di-sviluppo-edilizio/progetti-di-edilizia-universitaria/istituto. [Consultato il giorno: 23-07-2018]

[41] Debarati G-S, Philippe H, Regina B. Annual Disaster Statistical Review 2015: The Numbers and Trends. Brussels, Belgium: Université catholique de Louvain; 2016

[42] Lawton JH. What do species do in ecosystems? Oikos. 1994;**71**(3):367-374

[43] Costanza R, d'Arge R, de Groot R, Faber S, Grasso M, Hannon B, et al. The value of the world's ecosystem services and natural capital. Nature. 1997;**387**:253-260

[44] Dudley N, Higgins-Zogib L, MacKinnon K, Sandwith T, Stolton S. National Parks with benefits: How protecting the planet's biodiversity also provide ecosystem service. Solutions for a Sustainable and Desiderable Future. 2011;**2**(6):87-95

[45] Fischer B, Turner RK. Ecosystem services: Classification for valuation. Biological Conservation. 2008;**141**:1167-1169

[46] Ministero dell'ambiente e della tutela del territorio e del mare. Definizione del metodo per la classificazione e la quantificazione dei servizi ecosistemici in Italia; 2009

[47] Grêt-Regamey A, Brunner SH, Kienast F. Mountain ecosystem services: Who cares? International Mountain Research and Development. International Mountain Society; 2012;**32**:S23-S34

[48] Haida C, Ru¨disser J, Tappeiner U. Ecosystem services in mountain regions: Experts' perceptions. Regional Environmental Change. 2016;(16):1989-2004

[49] Massimo De Marchi RS. RS. La valutazione economica dei servizi ecosistemici e del paesaggio nel Parco Naturale Adamello Brenta. Valutazione Ambientale. 2012:54-60

[50] Casucci S, Liberatore P. Una valutazione economica dei danni causati dai disastri naturali [Online]. 2012. Available from: http://www.irpi.cnr.it/outreach/una-valutazione-economica-dei-danni-causati-dai-disastri-naturali/

[51] Marin G, Modica M. Socio-economic exposure to natural disasters. Environmental Impact Assessment Review. Elsevier Science. 2017;**64**:57-66

[52] Sterlacchini S, Zazzeri M, Genovese E, Modica M, Zoboli R. Flood Damage in Italy: Towards an Assessment Model of Reconstruction Costs. Vienna: EGU General Assembly; 2016;**11**

[53] Pellicani R, Westen CJV, Spilotro G. Assessing landslide exposure in areas with limited. Landslides. 2014;**11**:463-480

[54] Kaplan RS, Norton DP. The balanced scorecard—Measures that drive performance. Harvard Business Review. 1992;**70**(1):70-79

[55] Kaplan RS, Norton DP. The Balanced Scorecard: Translating Strategy into Action. Harvard, United States, Boston: Harvard Business School Press; 1996

[56] Kaplan RS, Norton DP. Strategy Maps: Converting Intangible Assets into Tangible Outcomes. Harvard, United States: Harvard Business School Press; 2004

[57] Amartya S. In: Adler MD, Posner EA, editors. The Discipline of Cost-Benefit Analysis, Cost-Benefit Analysis. Legal, Economic, and Philosophical Perspective. Chicago, United States: The University of Chicago Press; 2001

[58] Layard R, Glaister S. Cost-Benefit Analysis. Cambridge, United Kingdom: Cambridge University Press; 1994

[59] Dasgupta P, Mäler K-G. The environment: Environment and emerging development issues. In: Layard R, Glaister S, editors. Cost-Benefit Analysis. Cambridge, United Kingdom: Cambridge University Press; 1994

Chapter 7

Lifecycle Risk Modelling of Complex Projects

Matthew Cook and John P.T. Mo

Abstract

Large, complex and challenging engineering projects require extensive understanding and management of risk. How these risks are identified at the infancy of a project and subsequently mitigated throughout the project lifecycle is critical to successful delivery. Many projects begin with a comprehensive attempt to identify risks but lack the tools to manage and measure risk mitigation as the project progresses through the lifecycle causing the project to spiral out of control. This chapter outlines details of a risk model that uses a system within a system approach of identifying and segmenting risks. The model can then be analysed quantitatively and generate a visual lifecycle risk profile that allows the project team to monitor risks continuously in the project lifecycle. Furthermore, the use of a baseline or ideal project is proposed that is used as a measure of likely success against new projects.

Keywords: risk profile, enterprise modelling, project lifecycle assessment, supply chain risks

1. Introduction

Highly complex platform systems, such as ships, aircraft and land vehicles, present enormous technical and financial project challenges, and these include modifications and enhancements to their systems during their long service life. Many project management decisions are injudicious because they are made without a clear understanding of key risks and their consequences. This leads to budget overruns, schedule delays, system failures and ultimately disgruntled customers [1]. When faced with managing complex projects, a strategy many engineering organisations tend to adopt is the development of their own bespoke risk handling methods which attempt to control project failures with varying degrees of success [2].

According to the international standard on risk management ISO 31000, risk is the possibility of an undesirable event happening [3]. This definition is made up of two aspects, the 'severity' of the unpleasant something, and the probability or likelihood of this something actually 'happening'. When undertaking extensive, highly complex and challenging projects it is essential that any large organisation identify and manage all risks that could preclude success.

Unfortunately, in reality it is all too common in industry for risk to be treated as an afterthought and even in some cases seen as a box ticking exercise. It is crucial that a strategy is developed for identifying, handling and mitigating risks to ensure the success of the project. How well this is achieved can make or break the project and even the organisation.

In some organisations, the use of risk management and analysis tools may be conducted by dedicated risk engineers who are trained in such practice. However, in doing their work, risk engineers still rely on either project managers or engineers working on the project to provide appropriate information for risk assessment. While these individuals (e.g. project managers and engineers) are no doubt well aware of possible risks relating to their project, capturing information about the risks is often subject to influences and consideration of the relationships among parties in the project. Consequently, outcomes can vary significantly among different groups [4]. It is clear that a system to capture and analyse risks in a new project either prior to commencement or at early the stages of the project is desirable.

Risks in large engineering projects can come from many sources including uncertainties in the work which can influence and determine cost and time of execution. Essentially, every activity in the project has varying degrees of risk. Traditionally, the focus is on reliability, availability, maintainability and supportability (RAMS) [5]. Engineering design professionals usually use methods such as failure mode, effects & critical analysis, fault tree analysis and event tree analysis to assess the performance in a quantitative value that is basically an indicator of risks [6]. Modarres [7] went further to identify, rank and predict contributors to risk. Modarres calculated probabilistic risk for different scenarios and offered some interesting methods of presenting risk in graphical forms. This work illustrated ways of quantifying risks and hence the possibility of ranking accordingly. Ayyub [8] used a number of real-life examples to illustrate ways risk data can be manipulated or partially used to achieve useful outcomes. Claypool et al. [9] conducted surveys of risk management techniques with 110 managers, who believed there was room for improvement. They particularly highlighted that little work has been conducted into reducing risk in the supply chain which large scale engineering projects depend heavily upon.

Abi-Karam [10] studied design-build type of construction projects and identified the risks in the proposal, pricing, project schedule, performance measures, contractual liability and safety areas. These risks should be identified as thoroughly as possible and managed continuously even beyond project completion.

Through their lifecycle management strategy, many organisations operate a risk and opportunity management plan (ROMP) for business units, projects and functions. Risks and opportunities are inherent in all project and business activities. Therefore, it is the responsibility of staff to continuously manage these risks and to promote and realise any opportunities. It is important however, to recognise that the aim of ROMP is not to eliminate risk totally but to provide a systematic means to proactively control and direct the business in regard to mitigating risks and promoting opportunities, thus creating and protecting business value. It is noted that the content of risk level on a project, determined as a result of mapping to the standard ISO risk matrix [2], can vary between companies.

The common tool for risk management in large engineering organisations is a risk register that records the opinion of project managers, engineers and other key staff involved in the project. The process of generating this register is often very subjective, and the assessment team may include internal and external stakeholders such as customers and alliance partners who participate in workshops, brainstorming and project meetings.

Complex engineering projects are usually coordinated by systems engineering methodology. It can be seen that the systems engineering approach has a more comprehensive coverage of the project development context in relation to large complex projects using a systems engineering management plan (SEMP) [11]. A typical engineering project in the defence environment for example will follow a

SEMP process usually structured to include a number of mandatory stages and theoretical gates which need to be passed before the change can be progressed. One of the risk management core activities during the systems engineering project lifecycle process is incorporating risk identification, analysis and mitigations throughout the SEMP stages. Risk assessment of the whole lifecycle in the project development stage in many cases lacks project details/data and is not adequately addressed at the critical early stage.

This chapter examines the SEMP and investigates whether risk management of the project can be defined much earlier in the process. It focuses on developing a quantitative risk model that can identify risks, develop a risk profile that can be presented in a visual format and manage/track residual risks throughout a project's life cycle. The potential development of a relationship between the model and a risk burndown chart, offers a means of associating the identified risks with both their predicted financial and schedule impacts and what affect proposed mitigations will achieve [12]. The focus of this chapter is to assist large organisations to identify, visualise and manage risks throughout the lifecycle of a project. It should also be noted that some of the methods of calculation chosen are not mandatory and the proposed model has considered the need for flexibility to allow for alternative interpretations and weightings. The risk burndown chart would be an ideal tool for visualising how different strategies in the allocation of resources, financial investment, cash flow, technical challenge, etc. could affect risks [13]. A risk model is regarded as useful if it can identify key risks from quantitative data and suggests possible mitigation strategies. While many organisations already attend to highlighting risks with an array of tools, software and/or process methods, their calibre is often diminished by over-complexity and convoluted processes that are too involved. Hence, probably the most important characteristics of a risk model should be simplicity and ease of use.

2. Risk assessment using a system model

The main problem with a risk register is the lack of a structured methodology for identifying risks and a systematic analysis process to determine and develop mitigation strategies for the complete engineering lifecycle. A generic enterprise model is necessary to provide a quantitatively generated risk profile [14]. The model is not intended to compete or replace current risk theory, tools or processes already available but instead offer a novel and enhanced method for managing and importantly visualising risks throughout a project lifecycle.

The approach described in this chapter takes into account the fact that large complex projects have several distinct stages and can last for years. This consists of developing an understanding of the factors affecting outcomes based on the contractual environment involving key stakeholders in the industry, the customer and the community at large. The purpose of modelling is to develop a numerical indicator which can be used to ameliorate the current processes involved in understanding and managing risk throughout the project lifecycle due to changes of these factors.

In order to set some form of qualitative baseline which can then be used for both quantitative assessment and analysis, the 3PE model (People, Process, Product & Environment), described by Mo [15] (**Figure 1**) has been adopted to provide the system framework for investigating risks surrounding complex engineering projects [16].

In **Figure 1**, the physical elements of a system are shown as: the 'product' that is built from fundamental engineering sciences, this is the common view of most users and society in general. The 'product' is the tangible element that can usually give

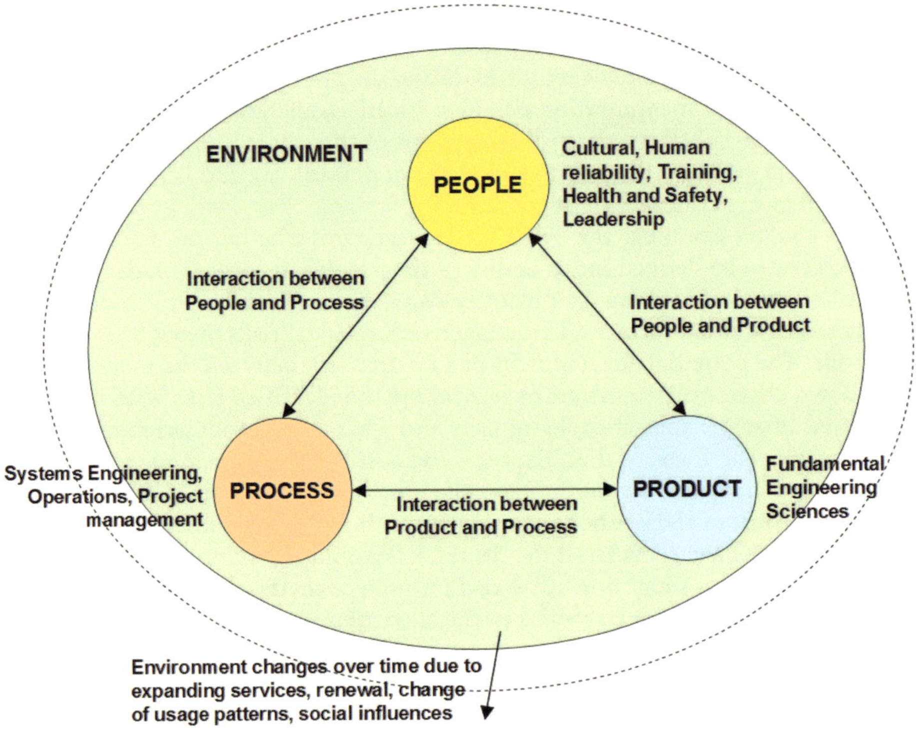

Figure 1.
Product process people environment (3PE) system model.

the 'touch-and-feel'. In commercial sense, this is what the customer feels they are paying for.

Not everyone realises that the 'people' element is an integral part of the system. The element 'people' from the system's point of view is not limited to the user. It includes all human participants who are involved, one way or another, to enable successful operation of the system to achieve its goals and applies to systems of any nature. In engineering projects involving design and build, the people are engineers, suppliers, technicians, managers, directors, stakeholders and customers from all organisations having an interest in the project.

To operate the 'product' properly, a set of procedures, i.e. 'process', should be defined and followed by everyone. The principles and practices of different people and organisations need to be synchronised and shared. This is the realisation of 'practices' that are accumulated, engineered and designed to generate knowledge of how to go about doing something and ensure success. A defined set of procedures not only allows the 'people' (remember there could be many people) to synchronise with the reactions of the system at different inputs during operation, but also ensures the system can overcome the challenges in operation.

Needless to say, these elements are interacting among themselves as shown by the double arrows between the elements in **Figure 1**. Without these interactions, the 'product' is not used by 'people', the 'people' do not follow the 'process' and the reaction of the 'product' is unpredictable. The outcome is obviously unsuccessful operational performance.

On top of this, while the three elements are working and interacting among themselves as a system, they co-exist within an 'environment'. If the 'environment' is within the expectation of the system, the elements in the system can work and interact correctly and produce good system performance. On the contrary, if the 'environment' has changed to beyond extreme conditions, the system will fail. To overcome this problem, the system has to be changed, i.e. some or all of the elements have to be changed to adapt to the new 'environment'. If nothing is done to the system while the 'environment' has changed, the system can potentially become out-of-date and/or obsolete.

In summary, the main elements in the 3PE model are people, process and product, which are located within an environment [17]. For each of the elements, a list of generic risks, more or less common to all projects was developed which were subsequently reconfigured into baseline questions. Once the model is defined, risk elements can be assigned and assessed. The risk indicator of a project can be estimated from the 3PE model as a normalised distribution of risk, in this project it is denoted by $N(\mu_j, \sigma_j)$, where j is a particular project.

In order to evolve the risk model further, the theory of generating a percentage of success for a given project has been used. The hypothesis being that a 'desirable' project, would have minimal risk that could be easily mitigated and has a percentage of success which can be established as the benchmark (i.e. high percentage of success). The 'desirable' project is defined as a distribution $N(\mu_d, \sigma_d)$.

To calculate the risk of not achieving project success, the differential distribution will show the risk of the project in relation to the 'desirable' project. The mean and standard deviation can be calculated using equations:

$$\mu_F = \mu_j - \mu_d \tag{1}$$

$$\sigma_F = \sqrt{\frac{\sigma_j^2 + \sigma_d^2}{2}} \tag{2}$$

The risk indicator at time of measurement is then defined as:

$$F = \left\{\Pr\left(\mu_j - \mu_d\right) > 0\right\} \tag{3}$$

The 'desirable' project is a reference only, it may still have some risks, but they should be acceptable and manageable. Eq. (3) is showing the probability of any project having a probability of failure that is greater than the 'desirable' project.

3. Case studies

Three projects were chosen to illustrate the methodology of calculating the risk of a project. These projects were or are being executed in the defence environment but the general principles of computing risk indicator F applies to all large complex projects.

PL1. This project was completed on budget and schedule with successful commissioning on site and acceptance by the customer. This project is considered medium size and combined OEM equipment and a customised installation. The normalised distribution of risk in this project is denoted by $N(\mu_1, \sigma_1)$.

PL2. This project was conducted in an alliance between two large defence contractors and the government. In this project, the highest risk was the need for a new technology to be developed by one of the collaborating companies. Since the new technology has not been substantially applied in the mission environment, it is considered to be a significant risk surrounding the project.

There was also considerable risk introduced by forming an alliance [18]. The normalised distribution of risk in this project is denoted by $N(\mu_2, \sigma_2)$.

PL3. This project relates to the design, manufacture and installation of an enhancement for a specific class of naval ships. The size of the project is considered medium. Since the enhancement is not a complex item, the project is considered manageable as the design, fabrication and installation is to be fully controlled by the individual organisation. The normalised distribution of risk in this project is denoted by $N(\mu_3, \sigma_3)$.

Risks were extracted from each of these projects, with over 150 risks being identified. The compiled risks were then analysed for repeats and commonality within each of the elements and a list of common risks across the three projects established. To help focus the research in developing quantification methodology, 10 risks from each of the 3P categories (total 30 risks) were selected based on their generic nature and applicability to the majority of past projects. Project managers, engineers and key staff members who were involved in these projects were asked to rate these risks on a scale of 0–10, where 0 means no risk at all and 10 means extremely high risk. The summarised results of their ratings are shown in **Table 1**.

It should be noted that a higher value in **Table 1** indicates a higher risk ranking.

The three projects can be represented by a normal distribution profile as shown in **Figure 2**. The graph in **Figure 2** can be interpreted as follows: Using PL2 as the example. The normal distribution of PL2 is represented as $N(7.4000, 2.2084)$. According to statistical analysis, the probability of failure is the area under the curve. However, the whole PL2 distribution is in the positive side of the profile, clearly this does not mean that the probability of failure of PL2 is almost 100%. The level of risk needs to be indicated as a relative value to some reference point or project(s).

	PL1		*PL2*		*PL3*	
	Mean	*Std. dev.*	*Mean*	*Std. dev.*	*Mean*	*Std. dev.*
Product	6.8143	2.3676	7.5286	2.4800	6.2714	2.5929
Process	6.5929	2.5101	7.0214	2.1707	6.6143	2.4276
People	7.1786	2.2321	7.6500	1.9413	7.2714	2.0736
3P combined	6.8619	2.3727	7.4000	2.2084	6.7190	2.4488

Table 1.
Data analysis for three projects.

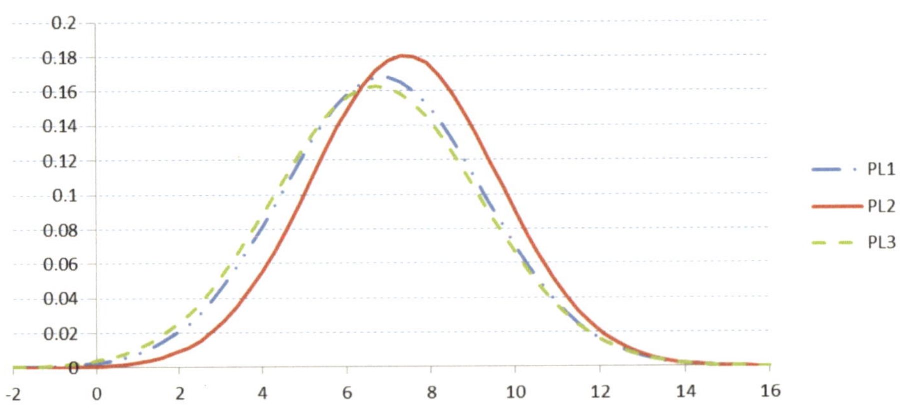

Figure 2.
Risk survey outcomes of 3P combined expressed in the form of a normal distribution.

From the perceived understanding of the nature of the three projects, it is generally agreed that serious challenges relating to PL2 need to be overcome and it is therefore considered a 'risky' project. PL1 has actually been completed and generally considered a success, while the PL3 project is clear in scope and is found to sit somewhere between the two. It can be seen in **Figure 2** that PL2 has a higher risk level than the other two projects (skewed more to the right). In order to develop the risk model further, the idea of comparing to a 'Desirable' or 'Ideal' project is explored.

As previously mentioned, the PL1 project is generally considered a successful project (PL1 was delivered within budget and on time. It was also viewed favourably by the customer as addressing their original need, hence why it is considered a successful project). It can therefore be judged that its data results must in some way align towards a 'desirable' project (i.e. minimal risk). To illustrate the computational process, the 'desirable' project could be defined from PL1 as an improvement of one ranking better than the PL1 data, i.e. minus one rating value. The outcome of this improvement gives a distribution $N(\mu_d, \sigma_d)$ as shown in **Table 2**. It should be noted that other methods of setting the benchmark 'desirable' project can be used, for example, survey a special expert group or find the 'reasonably smooth running' project. However, in the context of computing a risk indicator/gauge for these projects, the outcome does not affect the methodology discussed in this chapter.

Applying Eqs. (1) to (3), the differential risk profile of the three projects can now be calculated as shown in **Table 3**. The data can be represented by the risk profiles as shown in **Figure 3**. This set of risk profiles can be interpreted more rationally.

In **Figure 3**, using PL2 as the example, the differential normal distribution of PL2 against the 'desirable' project is computed by Eq. (1) and Eq. (2) as $N(1.5381, 4.5260)$. According to Eq. (3), the probability of failure is the area under the curve at the right-hand side of the y-axis (i.e. $x = 0$). In this area, the risk rating of PL2 is higher than the corresponding risk rating of the 'desirable' project. The area is calculated as 63.3%. Likewise, the probability of failure of PL1 and PL3 can be computed as 58.6 and 57.3% respectively. Please note that since the 'desirable' project in this case is generated from 'improving' upon the values of the best of the three projects, all projects will obviously have a riskier profile than the 'desirable'

	Mean	*Std. dev.*
Product	5.8143	2.3676
Process	5.5929	2.5101
People	6.1786	2.2321
3P combined	5.8619	3.9507

Table 2.
Data for the 'desirable' projects.

	Mean	*Std. dev.*
PL1	1.0000	4.6084
PL2	1.5381	4.5260
PL3	0.8571	4.6481

Table 3.
Differential risk profile data for the three case projects.

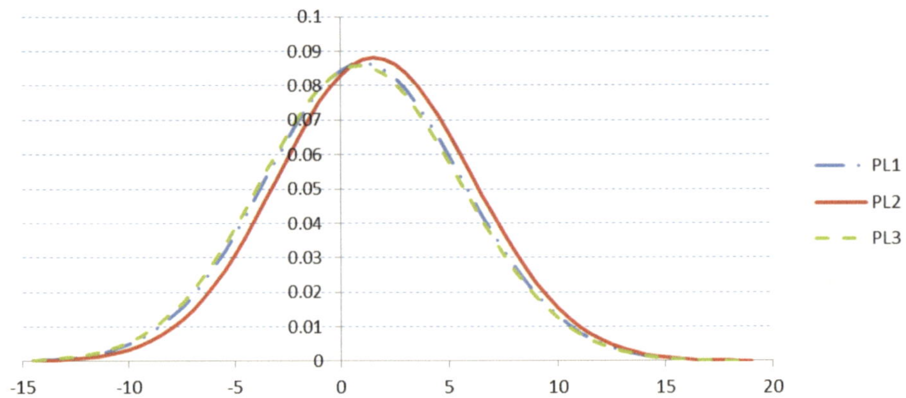

Figure 3.
Risk profile expressed as a normal distribution.

project. It is theoretically possible to manage a project better than the 'desirable' project. In that case, the probability of failure can be reduced significantly, approaching but not reaching zero.

4. Lifecycle risk assessment

Following the Systems Engineering V lifecycle [19] is an important decision towards successful management of the system development process. However, each of the stages in the systems engineering cycle is still lengthy and laden with risks. The systems engineering approach manages projects by a process of decisions and/or milestones.

To maintain the momentum and ascertain the right direction of development, it is necessary to impose routine checks throughout the forward branch of the V cycle with different level reviews. **Figure 4** shows a typical systems engineering management plan time line with different stages and reviews (sometimes known as 'milestones') clearly defined. Technically, the milestones serve as 'gates' that control the flow of the project. If the design or progress is not acceptable, the systems team is not allowed to start work towards the next review, hence it is a 'go or no-go' gate.

Please note that these reviews are the minimum number of checks that should be installed in the forward systems design cycle. More frequent, and less formal reviews can happen at any time and anywhere in the duration.

Initially **Figure 4** looks complicated but it is basically a representation of all project activities in parallel. Starting from the top bar which represents the activities to prepare a site, this line of activities will go through three reviews: system requirements review (SRR), preliminary design review (PDR), and critical design review (CDR). The sub-system 'user system' on the next bar will go through four reviews: concept design review (CoDR), SRR, PDR, CDR. The sub-system of 'integration and test facility' are complimentary outcomes of the next few design activities and it will be synchronised with those activities according to the red links. The most important sub-systems are represented by the next three bars, in this case: high resolution receiver, power system and signal processing. These sub-systems require a final pre-production review (PRR) before going into actual manufacture. Two other reviews important to the cycle are the site acceptance test (SAT) and test readiness review (TRR). Both these tests will involve the customer and reference verification to the user requirements.

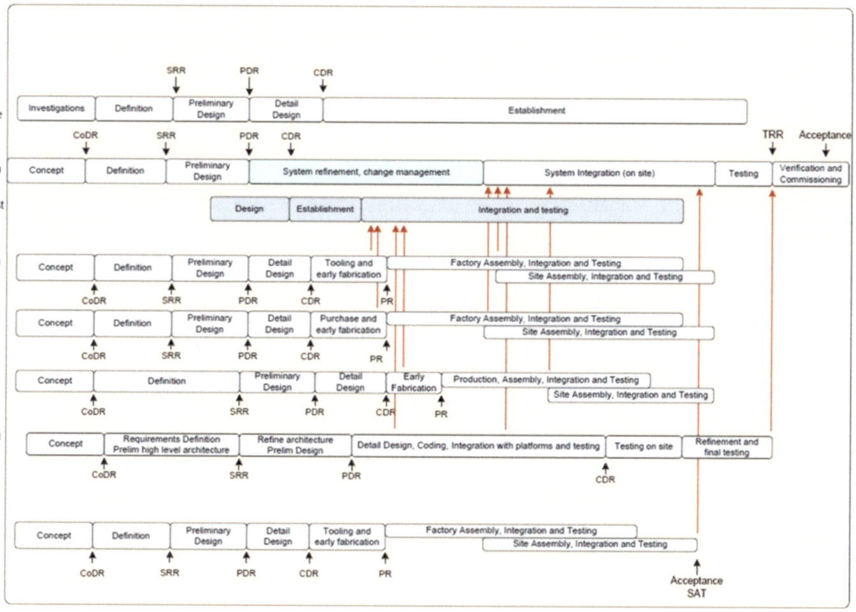

Figure 4.
Systems engineering management plan (for a communication system design) forward cycle with reviews.

The estimated success probability of the project at milestone i can be estimated using equation:

$$F_i = \left\{ \Pr\left(\mu_{ji} - \mu_d \right) > 0 \right\} \tag{4}$$

Hence, the estimated failure probability in this milestone is given by:

$$R_i = 1 - F_i \tag{5}$$

To determine the change of probability in the lifecycle, each of the process stages (in theory) should reduce or mitigate project known risks [20]. Therefore, if the project is progressing though the lifecycle as forecast, the risks of previous stages should be mitigated or resolved to the expected level. Post each stage, the project risk then comprises of the risks in achieving the remaining milestones. A way of representing this mathematically, is by the logic of compound events that are in series. If there are M milestones in the project, the project probability of success of the kth milestones is the product of probability of success of all milestones from milestone k:

$$R_k = \prod_{i=k}^{M} R_i \tag{6}$$

Therefore, the risk of the project at milestone k is given by:

$$F_k = 1 - \prod_{i=k}^{M} R_i \tag{7}$$

With the formula for estimating the risk of a project at the different milestones established, the next challenge is to determine how the success index and standard deviation of a complex project can be reasonably estimated. This leads finally to the calculation for the overall project success index, which is essentially a means of accurately estimating/forecasting based on a known range of required activities within the project.

As defined earlier, the 3PE model elements are people, process and product, located within an environment. A list of activities for each element and relevant interactions between these elements can be established and thus a holistic view of the project can be formulated. The organisations systems engineering team can then estimate the success index levels separately for each element. The sum of risks for each of the project stages can then be presented as a risk level or profile of the project. However, it should be noted that this profile represents a snapshot of the risks in the system at a single moment in time.

Any activity in a project can be assessed by combining the 3PE indices. The indices of activity j in milestone i can be denoted by μ_{ji}, and σ_{ij}. The estimated success index is now combined with both the desirable project mean index and the standard deviation. If normal distribution for all activity indices in milestone i is assumed, and there are α activities in the milestone, the resulting combined distribution of the milestone i is $N(s_{ji}, \sigma_{ji})$, where

$$s_{ri} = \frac{\sum_{j=1}^{\alpha} s_{rij}}{\alpha} \tag{8}$$

$$\sigma_{ri} = \sqrt{\frac{\sum_{j=1}^{\alpha} \sigma_{rij}^2}{\alpha}} \tag{9}$$

The probability of success is calculated with Eq. (4) and the overall or holistic project risk for a milestone can then be computed by Eq. (7). Again, it is worth highlighting that the identified risks of previous stages should be resolved and/or reduced (preferable approaching zero) as the project progresses through the lifecycle. It is then possible to plot risk levels against the project stages to visually highlight the reduction.

5. Lifecycle risk assessment: worked example

For this worked example, the theoretical risk profiling methodology is applied to a logistics improvement project as an illustration of the process. The company employs approx. 3000 staff at the time of writing and was undergoing a series of significant changes. In order to stay competitive, the organisation is developing a transportation network that will support its just-in-time (JIT) supply chain. However, to apply Eqs. (3) and (4), a conceptual desirable project is necessary. From the authors' previous experience on similar process improvement projects, it is desirable to set the mean success index and its standard deviation at s_d = 4.5, and σ_d = 0.5 respectively. It should be noted once again that the desirable project values used in this example are for illustration purposes only. Other methods of setting the benchmark or desirable project can be used as explained earlier. The scale used in this research is linear from 0 to 10 with 0 being no chance of success (sure failure) to 10 (being sure success).

After integrating the project's SEMP, the lifecycle process can be divided into six stages: (1) plan, (2) define, (3) preliminary design, (4) detailed design, (5) build, (6) deploy and close. At each milestone of the process, certain activities can be identified and these are aligned under the 3PE model elements as shown in **Table 4**. Each of the activities within the 3PE elements were assessed using a three-point estimate by the systems engineer of this project. The scale is the same as that used for the desirable project.

DOI: http://dx.doi.org/10.5772/intechopen.82273

	Plan	Define	Preliminary design	Detailed design	Build	Deploy and close
Product	• Statement of requirements	• Existing ineffective JIT • Transportation network • New technology • Scanning • Wireless communication	• Value stream analysis	• Software limitation • Subsystem detail design • Microsoft tools • Geocoding technology • CZAR rating system • Optimization technologies	• Implementation process	• Project management
Process	• Software system methodology • Root definition	• System of systems methodology • Strategic assumption surfacing and testing • Build variation and manual checking	• Define problem context (system requirements) • Hard system definition • System and user requirements • Environmental factors • Operational plan	• Dynamic operation environment, applicability to wide variety of situations, improved tracking of original orders, VSCM network design technology	• Key performance indicators, cost effectiveness, lifecycle costing	• Standard operating procedure
People	• Logistics Team	• Group formation	• Bargaining power of service providers	• Training	• Delivery	• Supply chain management
Process/ product	• Define context	• Supplier constraints • Loading plan • Input/ packaging • Vehicle profile • Equipment MGOA system	• Threat of substitute • Linear regression analysis	• Transportation standard compliance • Modelling and data parameters preparation	• System testing and validation • System function architecture verification • System synthesis development	• Process standardisation

	Plan	Define	Preliminary design	Detailed design	Build	Deploy and close
Process/ people	• Roles and responsibilities	• Declared assumptions • Synthesis	• Government policy • System dynamic • Business case approval	• Batch routing software • Scheduled network pooling, integration • Time-based schedule windows • Reports and business graphics • Data validation and scrubbing • Cubic calculation origin destination	• Ready to deploy documentation	• Extent of compliance
Product/ people	• Knowledge of people in position	• Project profile approval	• Establish baseline model principle • Baseline prediction and adjustments	• Communication user interface, Geographical data and sharing, Shipment information linear code reporting • Line haul route design	• System implementation	• System debugging and upgrade
Product/ people/ process	• CATWOE	• System support	• System breakdown structure • Identify subsystems and attributes • New transportation network overview • Individual city considerations	• Validation	• System performance indicators	• Extent of compliance

Table 4.
3PE model of logistics system improvement project.

By applying Eqs. (1) and (2) to the quantified 3PE elements, the numbers can be converted to normal distributions. As an illustration, the Process element is used to demonstrate the analytic process as shown in **Table 5**.

Using the means of the 3PE elements at each milestone and compared to the conceptual desirable project represented by a normal distribution $N(4.5, 0.5)$, the risk distribution of the SEMP cycle can be summarised in **Table 6**.

Combining the lifecycle risk of each 3PE element,, the risk levels at each milestone can be computed in **Table 7**.

The overall risk is plotted as a 'risk burn down' graph which highlights that initially at the planning stage of the project, there are many unknowns and uncertainties in the project (including those downstream) and the risk is considerably high. See **Figure 5**.

The risk computation as shown in **Table 5** indicates that some activities are risky, e.g. 'product' (in this case, implementation of the JIT transportation network

Milestone	Activity	Pess.	Nrm.	Opt.	s_{rij}	σ_{rij}	s_{ri}	σ_{ri}
Plan	Software system methodology	3	5	6	5.500	0.316	1.000	0.418
	Root definition	5	6	6				
Define	System of systems methodology	2	5	7	4.556	0.460	0.056	0.480
	Strategic assumption surfacing and testing	3	4	5				
	Build variation and manual checking	3	5	6				
Prel. design	Define problem context (system requirements)	3	5	6	4.333	0.422	−0.167	0.462
	Hard system definition	3	4	6				
	System and user requirements	3	5	6				
	Environmental factors	3	3	4				
	Operational plan	3	5	5				
Detail design	Dynamic operation environment	1	3	4	3.667	0.380	−0.833	0.444
	Applicability to wide variety of situations	3	5	6				
	Improved tracking of original orders	3	4	5				
	VSCM network design technology	2	3	4				
Build	Life cycle costing	5	7	8	5.667	0.715	1.167	0.617
	Alternatives evaluation	3	6	9				
	Logistics support analysis	3	5	6				
	Technical factors (operational effectiveness)	4	8	9				
	Key performance indicators	3	4	5				
	Economic factors (cost-effectiveness)	3	5	6				
Deploy and close	Standard operating procedure	3	5	6	4.833	0.707	0.333	0.612

Table 5.
Expanded 3PE process element.

	R_i *of element*	F_i of element	F_k of element at milestone k
Product			
Plan	0.7069	0.2931	0.6380
Define	0.9347	0.3393	0.4879
Preliminary design	0.7069	0.5329	0.4521
Detail design	0.7984	0.6271	0.2249
Build	0.9853	0.6326	0.0292
Deploy/close	0.9853	0.6380	0.0147
Process			
plan	0.9916	0.0084	0.9960
Define	0.5461	0.4585	0.9959
Preliminary design	0.3593	0.8055	0.9925
Detail design	0.0303	0.9941	0.9792
Build	0.9707	0.9943	0.3138
Deploy/close	0.7069	0.9960	0.2931
People			
plan	0.9650	0.0350	0.8297
Define	0.7069	0.3179	0.8235
Preliminary design	0.7069	0.5178	0.7503
Detail design	0.7069	0.6591	0.6468
Build	0.7069	0.7590	0.5003
Deploy/close	0.7069	0.8297	0.2931
Process/product			
plan	0.7069	0.2931	0.8141
Define	0.9475	0.3302	0.7371
Preliminary design	1.0000	0.3302	0.7225
Detail design	0.9768	0.3458	0.7225
Build	0.4019	0.7371	0.7159
Deploy/close	0.7069	0.8141	0.2931
Process/people			
plan	0.7069	0.2931	0.8039
Define	0.8277	0.4149	0.7225
Preliminary design	0.7968	0.5338	0.6648
Detail design	0.7731	0.6396	0.5792
Build	0.7069	0.7452	0.4558
Deploy/close	0.7699	0.8039	0.2301
Product/people			
plan	0.8227	0.1773	0.6123
Define	0.7699	0.3666	0.5287
Preliminary design	0.7872	0.5014	0.3879
Detail design	0.9795	0.5116	0.2224

	R_i *of element*	F_i of element	F_k of element at milestone k
Build	0.8175	0.6007	0.2061
Deploy/close	0.9711	0.6123	0.0289
Product/people/process (all three)			
plan	0.8227	0.1773	0.9987
Define	0.9968	0.1799	0.9984
Preliminary design	0.9788	0.1973	0.9984
Detail design	0.7069	0.4326	0.9984
Build	0.8227	0.5332	0.9977
Deploy/close	0.0027	0.9987	0.9973

Table 6.
Risk distributions at each SEMP stage.

	Product	Process	People	Process/ product	Process/ people	Product/ people	Product/ people/process	Overall
Plan	0.6380	0.9960	0.8297	0.8141	0.8039	0.6123	0.9987	0.8132
Define	0.4879	0.9959	0.8235	0.7371	0.7225	0.5287	0.9984	0.7563
Prel. design	0.4521	0.9925	0.7503	0.7225	0.6648	0.3879	0.9984	0.7098
Detail design	0.2249	0.9792	0.6468	0.7225	0.5792	0.2224	0.9984	0.6248
Build	0.0292	0.3138	0.5003	0.7159	0.4558	0.2061	0.9977	0.4598
Deploy/ close	0.0147	0.2931	0.2931	0.2931	0.2301	0.0289	0.9973	0.3072

Table 7.
Overall project risk at different milestones.

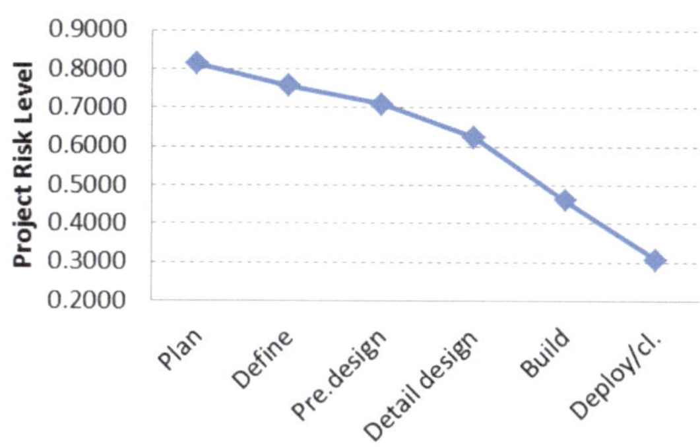

Figure 5.
Risk burndown plot.

system) at the 'deploy/close' stage has high probability of failure. The main uncertainty at this stage of the project was 'project management'. This is reflected by the fact that some resources have been redirected to other work when the JIT supply chain was close to completion.

The risk burndown plot can be used to verify the logical outcome of the project, i.e. as the project progresses through the lifecycle, many risks would have been resolved and thus the probability of failure or risk level should sequentially decrease. Failure of the project to follow this theory is an indication that the risks are not being managed as forecast.

6. Conclusion

This chapter combines the probability of success of every activity at each milestone of a project lifecycle based on a systems engineering process. The method uses an enterprise network model to study a manufacturing company's risk of undertaking a change project to re-design its logistics system in terms of planning, monitoring and validating of the network efficiency and criticality. Several key changes have been put in place and their risks in the system's lifecycle development are assessed within the enterprise network model to ensure greater probability of success.

The 3PE modelling framework provides a logical foundation for quantifying the risks of an engineering project. By assessing the expected level of achievable outcome in comparison against a 'desirable' project, this research has developed a novel method of generating a quantified risk indicator that can provide the basis for future planning improvements and hence the successful execution of complex engineering project. The use of a risk burndown plot allows an organisation to visualise the level of risk that has been burnt-down at each stage of lifecycle. In addition, the 3PE model provides a method for modelling different scenarios and the ability to assess their effectiveness at burning down risks.

Author details

Matthew Cook* and John P.T. Mo
RMIT University, Melbourne, Australia

*Address all correspondence to: cook.matthew@hotmail.com

References

[1] Cook M, Mo JPT. A systems approach to life cycle risk prediction for complex engineering projects. Cogent Engineering. March 2018;**5**(1):13. DOI: https://doi.org/10.1080/23311916.2018.1451289

[2] Baccarini D. The logical framework method for defining project success. Project Management Journal. 1999; **30**(4):25-32

[3] ISO. ISO31000:2009 Risk Management—Principle Guidelines. International Organization for Standardization; 2009. Available for purchase from: https://www.iso.org/standard/43170.html

[4] Wedley WC, Millet I. Modelling risk and uncertainty with the analytic hierarchy process. Journal of Multicriteria Decision Analysis. 2002;**11**: 97-107

[5] Barabadi A, Barabady J, Markeset T. Maintainability analysis considering time-dependent and time-independent covariates. Reliability Engineering & System Safety. 2011;**96**(1):210-217

[6] Markeset T, Kumar U. R & M and risk-analysis tools in product design, to reduce life-cycle cost and improve attractiveness. In: 2001 Proceedings Annual Reliability and Maintainability Symposium. Philadelphia, PA, USA: Pub.IEEE; 22–25 January 2001. pp. 116-122

[7] Modarres M. Risk Analysis in Engineering: Techniques Tools and Trends. 1st ed. Taylor & Francis; 2006 ISBN: 9781574447941

[8] Ayyub B. Risk Analysis in Engineering and Economics. 2nd ed. Chapman & Hall/CRC; 2014:640. ISBN: 978-1-46-651825-4

[9] Claypool E, Norman B, Needy KL. Design for supply chain: An analysis of key risk factors. Industrial Engineering and Management (Open Access). **4**(2):8. DOI: 10.4172/2169-0316.1000156

[10] Abi-Karam T. Managing risk in design-build. AACE International Transactions. 2001;**CD71**:CDR07.1-5

[11] Rodriguez TS. Systems Engineering Management Plans. Sandia Report SAND2009-7836. Sandia National Laboratories; October 2009. p. 30

[12] Cook M, Mo JPT. Strategic risk analysis of complex engineering system upgrade. European Scientific Journal. Oct, 2015;**2**:64-80

[13] Flood RL, Jackson MC. Creative Problem Solving: Total Systems Intervention. Chichester, New York: Wiley; 1991. p. 250. ISBN 0471930520

[14] Cete M, Yomralioglu T. Re-engineering of Turkish land administration. Survey Review. 2013; **45**(330):197-206

[15] Mo JPT. Performance assessment of product service system from system architecture perspectives. Advances in Decision Sciences. 2012;**2012**:19. article ID 640601

[16] Mo JPT, Richardson AC. Enterprise modelling of isolated support system. International Journal of Agile Systems and Management. 2017;**10**(3/4):341-371

[17] Mo JPT. The use of GERAM for design of a virtual enterprise for a ship maintenance consortium. In: Saha P, editor. Handbook of Enterprise Systems Architecture in Practice. Information Science Reference (IGI Global); 2007. pp. 351-365. ISBN 978-1-59904-189-6

[18] Cook M, Mo JPT. An investigation into the risk of forming an alliance. In: Transdisciplinary Engineering Methods for Social Innovation of Industry 4.0. Pub. IOS; 2018;7:1135-1144

[19] Mo JPT, Lorchirachoonkul W. Lifecycle design and support of intelligent web-based service systems. International Journal of Agile Systems and Management. 2016;**9**(2):135-153

[20] Mo JPT, Cook M. Quantitative lifecycle risk analysis of the development of a just-in-time transportation network system. Advanced Engineering Informatics. April 2018;**36**:76-85

Chapter 8

Risk Analysis Related to Cost and Schedule for a Bridge Construction Project

Rafiq M. Choudhry

Abstract

The construction sector is subject to more risk than many other sectors. Managing risk is the hottest topic of discussion for engineers within the construction sector. It is difficult to imagine managing of projects without risk management in construction. Risk management is concerned with risk management planning, identification, analysis, responses, monitoring and controlling project risk. Risk analysis is an evaluative process that establishes the magnitude of risks on projects. This work is planned to identify and analyze risks in the construction of a bridge project. The data are collected through a survey approach by administering a questionnaire. Professionals involved in the construction of bridges identify the project risks. A case study is utilized to determine the impact of cost and schedule risks. The analysis is carried out using the Monte Carlo simulation. The findings of the Monte Carlo simulation are compared with the actual times and costs of the case-study project. The results show the actual times and costs fell within the expected distribution of the simulation. The results indicate that risk analysis is helpful in managing costs and schedule risks. Additionally, this work documents guidelines for risk analysis.

Keywords: bridge project, cost risk, schedule risk, risk analysis, risk management, Monte Carlo simulation, risk guidelines

1. Introduction

Risk is the chance of something happening that has an impact upon objectives. It is measured in terms of consequences and likelihood. Probability, likelihood, and chance are used synonymously, as also are consequences and impact. Everyone struggles to understand and deal with many risk situations—from a domestic to corporate level, personal to national level, activity to project level. Management of projects can be improved by raising awareness about risks, and then implementing formal processes to deal with them. Construction projects are fundamentally susceptible to risks. Projects can be successful if managers plan for risks—planning, identifying, analyzing, and providing response for undesirable events that can occur. Choudhry and Iqbal [1] reveal that risk management is a new area in the construction sector and is attaining importance in the construction industry. The application of systematic risk management system is necessary for managing project risks [1]. Risk management can be applied successfully by identifying the risk sources connected with activities of the project. These risks are quantified in terms of likelihood and impact. Akintoye and Macleod [2] revealed that risk affects the performance, quality, budget and productivity of projects in construction. The strategy is to decrease the probability and impact of a risk [3]. Dikemen et al. [4] defined risk management as a systematic procedure of controlling risk. Choudhry and Iqbal [1] defined risk management as a stepwise process comprising on identification, analysis and risk response. Other researchers defined risk analysis as the procedure of evaluating identified risk and opportunities for their magnitude to proceed for a matching response in the light of limited funds.

Projects related to construction are complicated because they contain a range of human and non-human factors. These projects are started in complicated and vibrant environment resulting in high uncertainty, which are multifaceted by challenging time restraints [5]. Identifying and analyzing prospective risks can increase effective completion of the project. Risk management offers an opportunity for project stakeholders to review the project through a collective dialog, to recognize better and evaluate the prospective problems and then formulate a suitable response [6].

Various methods and models are developed by investigators to analyze risks. A decision support framework called as Advanced Programmatic Risk Analysis and Management Model (APRAM) is useful for risk management [7, 8]. Nasir et al. [9] devised a schedule risk model called as Evaluating Risk in Construction—Schedule Model (ERIC-S) that estimates the pessimistic and optimistic duration of activities. Ökmen and Öztas [10] proposed the Construction Schedule Risk Analysis Model (CSRAM) that evaluates schedule networks under uncertainty when duration of activities and risk factors are correlated. All these models evaluate either the time schedule risks, cost risks or both.

Risk management is vital in construction to minimize losses and improving profitability [2]. Williams et al. [11] proposed a method of risk management. Complicated projects, such as metro rail contains risks from the feasibility stage, construction to commissioning. Risks in heavy construction have a direct impact on the cost, schedule and performance. Reilly and Brown [12] reveal that infrastructure underground metro is inherently complex project having uncertain and variable ground situations. For these kinds of projects, it is vital to identify risk as early as possible in the project [12]. A risk management plan, if developed would ensure smooth attainment of project goals within given time, cost and quality. Moreover, it can safeguard better safety throughout the construction process and operative stage of the project.

Uncertainties in on-site and off-site project activities often result in the risk of delays and schedule overruns in construction projects. A risk analysis approach that assesses the integrating impact of uncertainties [13] show that growth in project size and work quantities intensifies pair and group interconnection of tasks within and between groups of on-site and off-site activities, resulting in lengthened completion times and deviations from project plans. Vu et al. [14] revealed that prolonged schedule delays have an extremely detrimental impact on a project's efficiency, cost and investment reputation.

Experienced experts are involved in analysis of risks qualitatively. Analyst needs to systematic and experienced to identify effectively internal or external risks. Mulholland and Christian [5] reveal that the decision maker makes the best use of experience and information in identifying risks. Akintoye and Macleod [2] reveal that an individual's attitudes, beliefs, and judgment can influence risk perception. Many professionals find that risk identification and risk analysis require involvement of experts and advanced techniques [1].

Choudhry and Iqbal [1] reveal that formal risk management is a rare in the construction industry of Pakistan. The authors explained that contractors are not practicing risk management formally. Major barriers to risk management are non-adaptation of formal risk management system [1]. Projects in the construction industry suffer in terms of low productivity, cost overruns and poor quality due not conducting risks management [1, 2]. The country is confronting the trauma of bridge failures and loss of lives every year due to floods. There is a need to develop risk analysis guidelines to avoid bridge failures. This chapter is to identify risks and critically rank them that affect the performance of project time and cost. Monte Carlo (MC) simulation on a case study project determines that risk analysis is helpful in managing schedule and costs risks. Identifying and analyzing schedule and costs risks on bridge project, this work makes a unique contribution and provides an insight into the risk management concepts.

2. Method

This work focuses on risk analysis by including a case study bridge project. The research investigates the impact of risks on costs, schedule and suggest guideline for bridge projects. To collect, data, a questionnaire is designed based on the previous studies [1, 15]. The questionnaire includes questions related to respondent identification, years of experience and 37 risk factors. Among the 37 risk factors, 7 are adopted from Choudhry and Iqbal [1], 8 are from Masood and Choudhry [15], 11 from the pilot survey, and 11 are developed by the researcher. These 37 risk factors are divided into 7 categories: design risks, contractual risks, construction risks, management risks, financial risks, health and safety risks, and external risks.

A pilot study is performed with a panel comprising five professionals having over 20 years of experience in construction. The questionnaire is modified based upon the pilot study. Based on importance of impact on the bridge project performance, respondents ranked each risk factor on a 5-point Likert scale (5 = extra ordinary, 4 = major, 3 = moderate, 2 = minor, 1 = insignificant). The respondents comprised on managers and engineers involved with numerous bridge projects. The targeted population for this work included private and public sector clients, consultants, and contractors. These include around 7000 enterprises that are involved in bridge construction projects and are registered with the Pakistan Engineering Council.

According to Dillman et al. [16], a sample size of 61 is fine with ±10% sampling error and a 95% confidence level. The respondents are approached through e-mails and personal visits to construction sites. Overall, 100 surveys forms are distributed on 25 construction sites. The response rate for this survey is 77%, but only 69 are analyzed. Eight surveys forms are not filled properly and thus discarded. Black et al. [17] stated that a 30% response rate is satisfactory in construction. The composition of the respondents is 35% public clients, 10% private clients, 12% contractors and 43% consultants. Public clients own most of the bridge projects. A majority of respondents are civil engineers holding a bachelor's degree with over 16 years of experience. In addition, 25 interviews are conducted; one at each project. These interviews delivered valuable information about risk management and risk analysis guidelines.

The collected data are analyzed by using a software called as Statistical Package for the Social Sciences (SPSS). Statistical techniques such as preliminary analysis, internal consistency analysis, relative importance index, Pearson's product–moment correlation are used in the analysis.

In addition, a case study of a bridge project is documented to establish costs and time risk analysis. The researcher obtained assistance from the five-member expert panel (comprising on scheduling manager, project manager, resident engineer, construction manager, an academia) and Monte Carlo simulation to analyze risks on of the case study project. The panel members are having more than 20 years of experience in industry and academics. This panel identifies the risks relevant to the case study project and assigned probability to the risk factors. This panel assigned the probabilistic (optimistic, most likely, pessimistic) durations and costs in Pakistan Rupees (PKR). These probabilistic durations and costs permitted us to practice triangular distribution in Primavera Pertmaster. A 3 days' workshop is held with the attendance of all panel members. Their involvements for the risk analysis are documented.

3. Results and analysis

3.1 Tests for factor analysis

Kaiser-Mayer-Olkin (KMO) test and Bartlett's test of sphericity are carried out to check the suitability for factor analysis. The suitability of a sample in relations to the distribution of the data is checked by KMO test. Pallant [18] stated that KMO value should be more than 0.5. The researchers [18, 19] revealed that factor analysis is meaningless with an identity matrix. The tests showed that KMO value is 0.689 that is more than 0.5 and Bartlett's test of sphericity is large (chi-square value = 1626.4890 with small significance (p value < 0.001).

3.2 Analysis for internal consistency

Cronbach's alpha (α) is used to check the internal consistency in the items involved in each factor [20] and the minimum recommended value is $\alpha = 0.7$ [19]. Cronbach's α [19] also measures the reliability of all factors. Factor analysis shows that all 7 themes of the 37 factors had Cronbach's α ranging from 0.921 to 0.912, which means that all the variables are reliable [19]. For all 37 variables, the overall α is 0.917.

3.3 Relative importance index

Chan and Kumaraswamy [21] reveal that the mean and standard deviation of individual factor are not appropriate to decide the total ranking as they do not indicate any association among the factors. As a substitute, we calculate the weighted average for every factor and formerly divide them with the highest scale of the dimension. The researchers [19, 21, 22] indicate that this results in a relative importance index. Respondents provide their responses on a Likert scale about the standing of the 37 risks affecting the cost and schedule aims of the project. Shash [22] provided the formula for relative importance index as:

$$\text{Relative importance index (RII)} = \sum(aX) \times 100/5 \qquad (1)$$

where 'n' is the frequency of the responses; and 'N' is the total number of responses that gives $X = n/N$. Where 'a' is the constant that express weight specified to each response, ranging from 1 (insignificant) to 5 (extra ordinary).

The relative importance index categorized the seven risk factors in descending order as: financial risks (RII = 69.95), external risks (RII = 66.67), design risks

Risk category	Relative importance index (RII)
Financial risk	69.95
External risk	66.67
Design risk	66.28
Management risk	65.17
Construction risk	62.72
Contractual risk	59.42
Health and safety risk	53.82

Table 1.
RII of risk categories.

(RII = 66.28), management risks (RII = 65.17), construction risks (RII = 62.72), contractual risks (RII = 59.42), and health and safety risks (RII = 53.82). According to the results, financial risks are vital in affecting the cost and schedule aims of projects (see **Table 1**). The 2nd and 3rd most important risks are external risks and design risks.

Among the 37 factors, the highest 10 risk factors in order of importance are: unavailability of funds (RII = 85.80), financial failure of contractor (RII = 76.52), poor site management and supervision (RII = 74.20), inadequate site investigation (RII = 73.91), inadequate project planning (RII = 73.91), construction delays (RII = 73.62), unavailability of land and/or right of way for site access (RII = 72.17), defective work and or quality issue (RII = 71.88), financial delays (RII = 71.01), insufficient technology (RII = 69.86). These risk factors are important for clients, consultants and contractors. There is a need for an effective risk management system on construction projects. Health and safety risks are ranked at the bottom in the 37 factors. This indicates that clients and consultants do not demand from contractors to implement a proper health and safety management system. Management risks are rated with high importance. There is lack of construction management experts and only few institutions offer program in construction management. Small contractors generally do not hire qualified engineers unless it is mandatory by the client. There is a need for construction management and risk management education as well as research in the industry.

3.4 Pearson's product-moment correlation

The Pearson product-moment correlation ('r' Rho) is a measure of the degree of linear relationship among the variables. The correlation coefficient ('r' Rho) is any value between plus and minus and the sign (±) explains the direction of the relationship, either positive or negative. A positive coefficient means that the value of the variable increases with the increase in value of the other variable; or if one goes down, the other also reduces. A negative coefficient indicates that as one variable increases, the other decreases, and vice-versa. The absolute value of the coefficient indicates the strength of the correlation. A coefficient of $r = 0.50$ shows a robust degree of linear relationship than that of $r = 0.30$. A coefficient of zero ($r = 0.0$) shows the lack of a linear relationship and coefficients of $r = +1.0$ and $r = -1.0$ show a perfect linear relationship [19].

Table 2 shows the Pearson's correlations for the risk factor categories. The maximum coefficient (0.756) is between the construction and management risks, which is significant at the p value = 0.01. This indicates that numerous construction and management risks are correlated to each other and they are to be jointly addressed

Risk factor category	Financial	Contractual	Design	Safety	Management	Construction	External
Financial	1						
Contractual	.442**	1					
Design	.306*	.374**	1				
Safety	.098	.428**	.341**	1			
Management	.174	.445**	.374**	.366**	1		
Construction	.113	.380**	.250*	.459**	.756**	1	
External	.162	.290*	.399**	.373**	.430**	.605**	1

**Correlation is significant at 0.05 level.*
***Correlation is significant at 0.01 level.*

Table 2.
Pearson's product-moment correlation of risk factor categories.

with good risk management practices. There is another essential coefficient of 0.605 at significance p value = 0.01 between construction and external risks. External risks impact on project costs and schedule more than the construction risks (see **Table 1**). They are in fact the second most important risk factor category. A positive correlation of health and safety risks with construction (0.459) at a significance p value = 0.01 confirms the importance of health and safety on bridge projects. Higher rate of risks in construction indicate an increase in physical vulnerabilities. The health and safety risks are correlated positively with contractual risks (0.428) at a significance p value = 0.01, indicating improvement in health and safety in construction may reduce contractual and health and safety risks.

3.5 Bridge project: a case study

The case study project is a bridge construction in Islamabad that links the Islamabad highway with a residential community. The project is located in the capital city of Pakistan. It has the following features: (a) bridge total length 166 m (544.8 ft), (b) constructed over a river with an annual peak discharge of 11,170 cusecs, (c) 56 piles of diameter 762 mm (2.5 ft) and abutment piles 15.24 m (50 feet) deep, (d) 4 spans, (e) pier piles 9.14 m (30 feet) deep, (f) 12 pile caps, (g) 4 abutment walls, (h) 2 abutments, (i) 12 piers, (j) 6 transoms or cross-beams, (k) 24 precast girders of 44.09 m (144.66 ft), (l) 14.32 m (47 feet) width of deck slab on one side, (m) 3.66 m (12 feet) length of approach slab on each side and, (n) asphalt 166.12 m (545 ft) long and the bridge is designed for 3 + 3 lanes of traffic.

A baseline work schedule is prepared for the project. The project has a base cost-estimate. Each activity in the schedule had its cost allocated. The allocation includes cost estimate for materials, equipment, labor, and overhead costs for each activity. The risks that are identified in the project are presented to the experts. The expert panel identifies specific risks to the case study project. These risks are loaded into the schedule to determine the impact on project schedule and cost. Primavera Pertmaster is used for risk analysis. The inputs to Pertmaster for the risk register are: (a) risk description, (b) risk ID number, (c) threat or opportunity, (d) effect of this risk on activity, (e) probability of occurrence, (f) type of risk e.g. schedule or cost, (g) distribution e.g. triangular, (h) correlation with other risk factors. The risk register (see **Figure 1**) is developed for the whole project.

Pertmaster uses Monte Carlo simulation for risk analysis. Monte Carlo simulation uses random independent variables to obtain solutions of problems. Lian

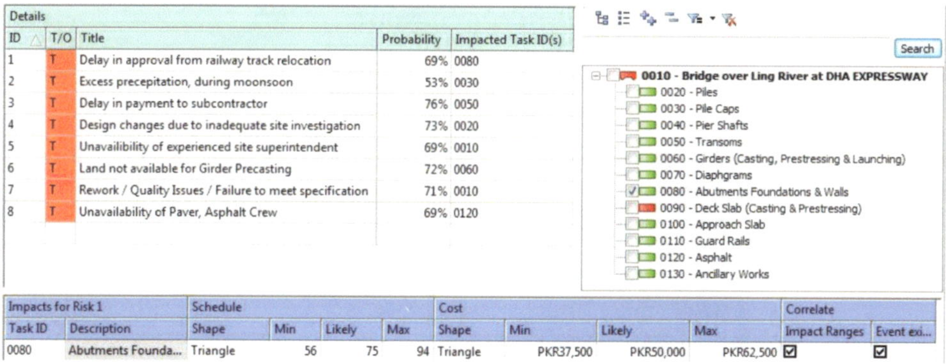

ID	T/O	Title	Probability	Impacted Task ID(s)
1	T	Delay in approval from railway track relocation	69%	0080
2	T	Excess precepitation, during moonsoon	53%	0030
3	T	Delay in payment to subcontractor	76%	0050
4	T	Design changes due to inadequate site investigation	73%	0020
5	T	Unavailibility of experienced site superintendent	69%	0010
6	T	Land not available for Girder Precasting	72%	0060
7	T	Rework / Quality Issues / Failure to meet specification	71%	0010
8	T	Unavailability of Paver, Asphalt Crew	69%	0120

Impacts for Risk 1		Schedule				Cost				Correlate	
Task ID	Description	Shape	Min	Likely	Max	Shape	Min	Likely	Max	Impact Ranges	Event exi...
0080	Abutments Founda...	Triangle	56	75	94	Triangle	PKR37,500	PKR50,000	PKR62,500	☑	☑

Figure 1.
Risk register of the project.

and Yen [23] reveal that Latin hypercube sampling and simple random number sampling are among the sampling techniques that are used with Monte Carlo simulations. This simple and elegant method delivered a means to solve equations with triangular probability distributions [24, 25]. Critical path is found and further calculations are documented with activities that are on the critical path. The time schedule loaded with costs and risks is analyzed. Real versus simulation outcomes are compared. A total of 1000 iterations are conducted for risk analysis.

The cumulative distribution for project cost, finish date, and duration are calculated with Monte Carlo simulations. The project duration (maximum = 792 days, minimum = 628 days, mean = 701 days) is displayed in **Figure 2**. The cumulative distribution for project duration and cost is determined. The results showed that the probability of finishing the project within the allotted time (628 days) is 4% and within the budget (PKR 129 million) is less than 1%. Terms P100 and P80 indicate the probabilities of 100 and 80% respectively. For instance, P80 shows that the project could be completed in 730 days with an amount of PKR 161 Million. There shall be 100% sure that the project would be completes in 792 days or even less with a cost amounting to PKR 166.5 Million or less.

The observations are performed for 5 months for the case study project. We have compared Pertmaster results with the actual completed activities. The researchers spent full time on-site to ensure extreme communication with the project implementation team. Documents are cautiously reviewed and are assimilated in analysis. On-site real data are equated with the simulation outcomes. The evaluation associated with schedule start dates is noted. For piling activity, actual start dates matched with the base line as well as with P80 and P100. For pier-shaft, actual start dates are between P80 and P100.

Finish dates are also compared for the case study project. Piles activity finishes between the forecasted dates of P80 and P100. Pile-cap activity finishes between the expected dates of P80 and P100. The 'pier-shaft' activity also accomplished 22 days before the P80 finish date. Transom activity finishes 19 days after the P100 completion. This indicates that the simulation results are precise as the activities are actually completing either within the predicted dates or within ±20 days.

For the case study, costs are compared that are important to the contract partners. The project cost is at all times important to the management team. Probabilistic cost calculation with the model is very precise as all the genuine costs fell within the P80 and P100. The project cost incurred up to the completion of transoms is PKR 72.8 Million, while that forecasted by simulation with 80% probability is PKR 69.8 Million. The evaluation is PKR 76.0 Million with 100% probability, indicating that the expected cost using Monte Carlo simulation is precise

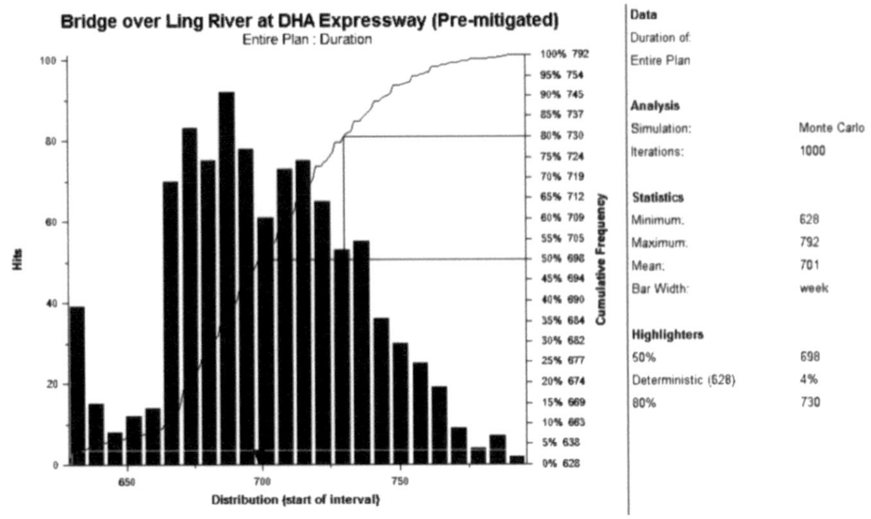

Figure 2.
Monte Carlo simulation results.

(P80 costs = 69.8 Million, Actual = 72.8 Million, P100 costs = 76.0 Million). The baseline cost of the project is only PKR 37.2 Million up to 'Transoms' construction that shows a cost overrun of 96%, portraying the absenteeism of monitoring and control of cost practices. This shows a clear requirement of risk management on bridge construction projects.

For the case study project, risk analysis shows that project management can obtain a fair idea of schedule and cost changes and variations. For the case study project, risks (see **Figure 1**) that affected the schedule and cost objectives are: (a) delay in approval from the regulatory authority i.e. delay in sanctioning relocation of the railway track, (b) unexpected weather i.e. excess rainfall during monsoon, (c) design variations i.e. design changes due to insufficient site investigation, (d) insufficient work space i.e. land not available for pre-casting of girders, (e) lack of technology i.e. breakdown in asphalt paving equipment, (f) unavailability of funds i.e. delay in payment to subcontractor, (g) unavailability of material i.e. quality issues and material failure to meet specifications.

3.6 Guidelines for risk analysis

The study advocates the succeeding guidelines for an effective risk analysis of any bridge project:

1. *Context development*: Developing the context for risk analysis is exceptionally vital as indicated in the 25 interviews documented. The expert panel emphasized the requirement for precise definition of the scope of the construction project; develop the project method statement, and conduct stakeholder analysis systematically. These points set the boundary for risk analysis as stressed by researchers [26]. Factors and variables contributing to project risks are required to be recognized.

2. *Identification of risks*: Tools and techniques such as checklists, historical data, brainstorming, and idea stimulating techniques may be employed [25, 26]. Nonetheless, risks are required to be identified and defined as is carried out

in the design of questionnaire. Help from expert panel should be sought in identifying risks. Choudhry and Iqbal [1] have documented the risks identification techniques and they may be adopted. Especially, risk related to time and cost is to be evaluated as it plays a major role in affecting the project performance.

3. *Quantifying risks*: The risk quantification is the most important process that requires skills, extensive experience and good judgment. In this process, one has to assess the probability of each risk [24, 25]. Next is to evaluate the impact of time or cost, or both. Generally, expert panels play a major role in calculating the probability of risks and their impact. The correlation of risks either positive or negative is addressed in this study. Lastly, the risk quantification decides, whether they have an effect on cost or duration, or both.

4. *Prepare cost-loaded schedule for the project*: Mubarak [27] revealed that the project baseline schedule needed to be prepared at the initial stage the project to measure the project's progress against it. Probabilistic or deterministic durations of time and costs of activities are to be estimated. The critical path needs to be determined based on the probabilistic durations. Project cost is determined based on the probabilistic costs of the activities' information. The comparison of actual duration and actual cost of activities is carried out with computed results and with the baseline.

5. *Schedule loading with risks*: When cost-loaded schedule is complete, the next step is to allocate risks as they are quantified with each of the project activities. These risks are generally documented in a risk register. The risk register contains all particulars of each risk for the project. From the risk register, relevant risks are assigned with the cost-loaded schedule.

6. *Running of Monte Carlo simulations*: The schedule loaded with risk is run by Monte Carlo simulations to calculate the impact. One needs to perform the Monte Carlo simulations by using software, for example, @Risk. Pertmaster is employed in this this research.

7. *Understand the output*: The results that are produced by Monte Carlo simulations are easy to comprehend. Outputs reflects the probability of meeting the time and costs. The P80 and P100 values represent 80 and 100% probability. They specify the values of time and cost with 80 and 100% confidence level. The results shows how much an activity is behind from its initial time and how much cost can overrun (see **Figure 2**).

This work reveals a systematic process to identify and quantify major risks related to construction and predominantly to bridge construction affecting cost and schedule of the project. All projects have their own special conditions; nonetheless, experts can acquire valuable evidence from the results as all projects have risks that need to be managed. Risks related to schedule create cost risk. The case study demonstrates with the help of Monte Carlo simulation that how schedule and cost risk can be analyzed and managed. The case study shows that understanding the probabilistic cost is vital to forecast long-term budgets. The risk management guidelines are documented from surveys, interviews and analysis. One can determine the probabilistic cost of project by adopting these guidelines.

4. Conclusions

This work has identified and ranked the critical risks threatening the performance of bridge construction projects and evaluated the consequence of risks on project time and cost. This work is planned to developed consciousness of project stakeholders in relation to risk analysis in the construction industry. The major risks concerning a bridge construction project are identified. After carrying out risk analysis, the major results of this work are examination of critical risks affecting project costs and schedule. Relative importance index of important risk factors is calculated. This exercise categorized risk that include 'financial risks', 'design risks', 'external risks', 'management risks', 'contractual risks', 'health & safety risks' and 'construction risks'. Financial risks are categorized at the top. The five highest ranked risk factors are 'financial failure of contractor', 'unavailability of funds', 'poor site management & supervision', 'inadequate project planning' and 'inadequate site investigation' among the 37 factors. Many risks are correlated and they need to be managed by applying management practices.

Schedule and costs risks are investigated in a case study project of a bridge construction. Real data of the bridge construction project is compared with simulation results after the risk analysis. Simulation findings are precisely correct and comparable to those really performed in relations to project duration and cost. In addition, guidelines for risk analysis are developed that can assist management in ascertaining possible risks on construction project of bridges. The research stresses that management is required to perform risks analysis after identifying prospective risks at the initial stage of bridge projects. Predicting risks can enable policymakers to detect areas of anxiety for project managers to take preemptive actions.

Although, each project in construction has its particular circumstances, project managers can acquire positive information from this study for their projects as the risks recognized for the construction are alike to risks in all sorts of projects across the world. The evidence delivered through this investigation can empower engineering professionals to safeguard that their projects advance efficiently without making unnecessary errors. This would be supportive to improve the execution of their projects. Even though there is some body of awareness in relation to management of risks, the formation of guidelines for risk management persisted to be vague. This research provided guidelines for risks management accompanying with heavy engineering construction in the construction sector. The research is important for project managers, academicians and professionals who are linked with heavy engineering construction and the construction sector in common.

Author details

Rafiq M. Choudhry
Construction Engineering and Management, Department of Civil Engineering, College of Engineering, Al Imam Mohammad Ibn Saud Islamic University, Riyadh, Kingdom of Saudi Arabia

*Address all correspondence to: rchoudhry@imamu.edu.sa; choudhry03@gmail.com

References

[1] Choudhry RM, Iqbal K. Identification of risk management system in construction industry in Pakistan. Journal of Management in Engineering. 2013;**29**:42-49

[2] Akintoye AS, Macleod MJ. Risk analysis and management in construction. International Journal of Project Management. 1997;**15**:31-38

[3] Zhang Y. Selecting risk response strategies considering project risk interdependence. International Journal of Project Management. 2016;**34**:819-830

[4] Dikmen I, Birgonul MT, Arikan AE. A critical review of risk management support tools. In: Khosrowshahi F, editor. 20th Annual ARCOM Conference. 2004. pp. 1145-1154

[5] Mulholland B, Christian J. Risk assessment in construction schedules. Journal of Construction Engineering and Management. 1999;**125**:8-15

[6] Iqbal S, Choudhry RM, Holschemacher K, et al. Risk management in construction projects. Technological and Economic Development of Economy. 2015;**21**:65-78

[7] Dillon RL, Pate-Cornell ME. APRAM: An advanced programmatic risk analysis method. International Journal of Technology, Policy and Management. 2001;**1**:47-55

[8] Imbeah W, Guikema S. Managing construction projects using the advanced programmatic risk analysis and management model. Journal of Construction Engineering and Management. 2009;**135**:772-781

[9] Nasir D, McCabe B, Hartono L. Evaluating risk in construction-schedule model (ERIC-S): Construction schedule risk model. Journal of Construction Engineering and Management. 2003;**129**:518-527

[10] Ökmen Ö, Öztaş A. Construction project network evaluation with correlated schedule risk analysis model. Journal of Construction Engineering and Management. 2008;**134**:49-63

[11] Williams RC, Walker JA, Dorofee AJ. Putting risk management into practice. IEEE Software. 1997;**14**:75-82

[12] Reilly J, Brown J. Tunnelling and underground space technology. In: International Tunneling Conference. Singapore: Pergamon; 2004. pp. 703-712

[13] Arashpour M, Wakefield R, Lee EWM, et al. Analysis of interacting uncertainties in on-site and off-site activities: Implications for hybrid construction. International Journal of Project Management. 2016;**34**:1393-1402

[14] Vu HA, Cu VH, Min LX, et al. Risk analysis of schedule delays in international highway projects in Vietnam using a structural equation model. Engineering Construction and Architectural Management. 2017;**24**:1018-1039

[15] Masood R, Choudhry RM. Identification of risks factors for construction contracting firms—Encompassing mitigation stance. In: Second International Conference on Construction in Developing Countries; Cairo, Egypt. 2010

[16] Dillman DA, Smyth JD, Christian LM. Internet, Phone, Mail and Mixed-Mode Surveys: The Tailored Design Method. New York: Wiley; 2014

[17] Black C, Akintoye A, Fitzgerald E. An analysis of success factors and

benefits of partnering in construction. International Journal of Project Management. 2000;**18**:423-434

[18] Pallant J. SPSS Survival Manual: A Step by Step Guide to Data Analysis Using SPSS for Windows. NSW: Allen Unwin; 2007

[19] Ghosh S, Jintanapakanont J. Identifying and assessing the critical risk factors in an underground rail project in Thailand: A factor analysis approach. International Journal of Project Management. 2004;**22**:633-643

[20] Carmines EG, Zeller RA. Reliability and Validity Assessment. Newbury Park, London: Sage Publication; 1979

[21] Chan DW, Kumaraswamy MM. A comparative study of causes of time overruns in Hong Kong construction projects. International Journal of Project Management. 1997;**15**:55-63

[22] Shash AA. Factors considered in tendering decisions by top UK contractors. Construction Management and Economics. 1993;**11**:111-118

[23] Lian Y, Yen BC. Comparison of risk calculation methods for a culvert. Journal of Hydraulic Engineering. 2003;**129**:140-152

[24] Love PED, Wang X, Sing C, et al. Determining the probability of project cost overruns. Journal of Construction Engineering and Management. 2013;**139**:321-330

[25] Choudhry RM, Aslam MA, Hinze JW, et al. Cost and schedule risk analysis of bridge construction in Pakistan: Establishing risk guidelines. Journal of Construction Engineering and Management. 2014;**140**:04014020

[26] Loosemore M, Raftery J, Reilly C, et al. Risk Management in Projects. London: Taylor & Francis; 2006

[27] Mubarak SA. Construction Project Scheduling and Control. 3rd ed. New Jersey: Wiley; 2015

Chapter 9

Pharmaceutical Projects: Walking along the Risk Management Line

Jordi Botet

Abstract

We manage risk so commonly (and unconsciously) in our everyday life that we tend to undervalue it. Risk management was officially introduced in the pharmaceutical world by the ICH guideline Q9 in 2005. Since then, it has been intensively used and, not infrequently, misused. Practice shows that risk assessment tools are often seen as an end in themselves, while such important things as brainstorming on the matter and getting to know the problem are underestimated. A pharmaceutical project provides a very good example of this: risk management is critical, but as there are many unknown factors, it has to be performed in a way that what really counts is understanding the problems we face. A pharmaceutical project has at least two actors, a pharmaceutical firm and an engineering company, possessing different backgrounds, and this often leads to different approaches. This may explain why risk management is not used as much as it should in pharmaceutical projects. Thus, this chapter considers a pharmaceutical project from the point of view of risk management.

Keywords: quality, pharmaceutical laboratory, engineering company, risk assessment tool, pharmaceutical quality system (PQS), life cycle, risk ranking and filtering (RRF), primary hazard analysis (PHA), biological agent

1. Introduction

Quality is not a matter of discussion. Quality is necessary for any manufactured product. Failures mean losses in terms of money, logistics, and prestige, and no industry can withstand these damages. Pharmaceutical products share this situation, but, besides, their lack of quality turns out to be a public health problem. This is why the authorities require that medicines meet their specifications and manufacturers have the responsibility of exclusively commercializing products possessing their purported quality.

The production of medicines is in fact composed of two different types of manufacture. Firstly, there are facilities that produce the ingredients, or active pharmaceutical ingredients (APIs), used in the preparation of the medicines. Secondly, we have those facilities, which combine the ingredients to obtain pharmaceutical forms. These latter are packed to get what we know as a drug, medicine, or pharmaceutical. This chapter focuses on this second type of manufacture, although we consider biotechnology too (in spite of producing substances, which are used as ingredients).

2. How to ensure quality

Taking for granted that medicines should always meet their specifications, we should find a reliable method for ensuring this. Unfortunately, guaranteeing quality is not an easy task, and this explains why over time different strategies have been applied.

The oldest one, which we might term as "analytical quality," proposed the analysis of the finished products as the tool for determining their appropriateness. This approach has many flaws, just to mention one, the quality problem is just detected when the product is already finished (and this makes corrective measures very difficult).

Then, in the middle of the twentieth century, "manufactured quality" was introduced. This new methodology, which led to the publication of the well-known GMP (good manufacturing practices), is based on the assumption that quality should be considered as another ingredient of the product (this was termed as "built-in quality").

More recently, already in the twenty-first century, the International Council for Harmonization (ICH) has extended this latter approach to the development stages of the product in order to attain a complete control of its life cycle. The logic underlying this method is very simple. How can you manufacture a quality that you have not designed previously? This is why we came to the present situation, which considers that quality is something that you should design (quality by design), construct (built-in quality), and supervise (process and product monitoring). In order to attain these goals, it is necessary to define and implement a policy of quality, which requires the establishment of a quality system. Although different systems are possible, in 2008, the ICH proposed a pharmaceutical quality system (PQS), specially developed for the pharmaceutical industry [1].

Even though the PQS focuses on pharmaceuticals, it is evident that if we are bound to produce quality products, the unit where we manufacture them should share this same approach for quality and here, the PQS provides very useful hints regarding the quality of pharmaceutical projects.

3. About the pharmaceutical project

A pharmaceutical project can be defined as a temporary effort undertaken with the aim of creating facilities that allow manufacturing medicines with the required quality and assurance. This effort is usually a quite difficult one, because it has to be carried out by different partners, which should work together, even if pertaining to different technical areas.

The partners who take part in a project belong normally to two groupings. On one side, the "client" (that is, the pharmaceutical laboratory or, maybe, laboratories) desires to possess new manufacturing premises or modify old existing ones. On the other side, the "supplier" usually consists of an engineering company that in fact coordinates different suppliers, extending from the providers of construction materials and associated services to the sellers of pharmaceutical equipment. These two parts, which to simplify matters we are going to name from now on as "laboratory" and "engineering," are usually sharply asymmetric. The laboratory in these matters, generally speaking, has a much more limited experience than the engineering, whereas the knowledge regarding pharmaceutical norms is often less developed in the engineering. The amount and importance of these differences can vary a lot, but here we have a possible source of problems that should not be overlooked and that makes very advisable tight control on a project.

4. Structure of a pharmaceutical quality system (PQS)

The paradigm of quality described by the ICH guidelines, Q8 to Q12 [1, 2–5], should also be applied to pharmaceutical projects, but taking into account the existing differences between manufacturing and managing a project. This is why not all the elements of a quality system devised for products can be applied into a pharmaceutical project. Let us analyze this (**Figure 1**).

The development and implementation of a PQS is the result of the definition of a quality policy by a laboratory. In order to put this policy into practice, the company writes a quality manual and associated documents that develop it in more detail. A quality manual should address the following topics:

1. The PQS is devised to ensure the application of GMP by the laboratory. As shown in **Figure 1**, the PQS covers the whole life cycle of a medicine, excluding the stage of development (this is the consequence of an old tradition of working in development centers and of the difficulty of applying the strict controls of GMP to a development laboratory).

2. Management responsibilities should be clearly stated. Practice shows that this is very important. As in old battles, the army can only win the war if there is a (good) chief leading it.

3. ICH Q10 describes two capacitors, intended to help in the task of reaching the objectives. The first one is risk management and we will discuss it in detail further on, as the main topic of this chapter. The second one is knowledge management and its intent is fighting the very pernicious practice that rules in more than one company: that is, people who get some knowledge on products

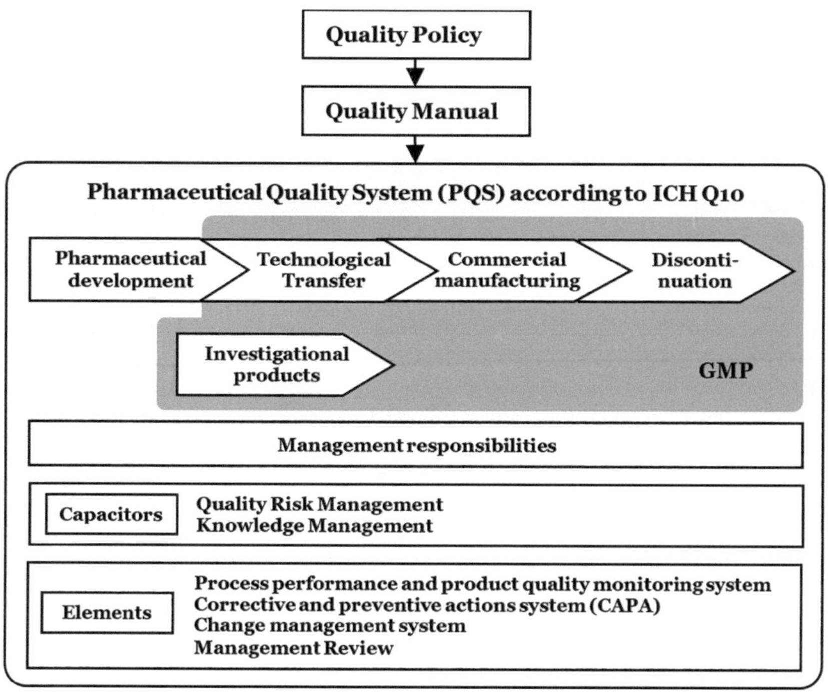

Figure 1.
Structure of the PQS (ICH Q10).

or processes do not inform the others, but keep it for them and, consequently, there is a continual loss of information.

4. The PQS contains four elements (**Figure 1**). All of them, albeit not in this form, are very important in a project.

5. Project life cycle

A life cycle approach is an important element for ensuring the global quality of a process. As mentioned before, quality has to be designed in order to be controlled. The term "designed quality," applied to a pharmaceutical project, means that objectives are well defined and that critical variables have been identified and quantified. This allows for a monitoring and evaluation of the project: if the critical variables are kept under control, then objectives will be met. Thus, the life cycle has to be considered as a chain of events that progressively increase the amount of information and build quality gradually.

As it can be seen (**Figure 2**), the life cycle of a project is composed of four phases, although only the first two (design and realization) concern what we have been calling pharmaceutical project and are managed by the two partners. The latter two phases (manufacture and discontinuation) are just the consequence of the former two and they belong exclusively to the laboratory. In fact, instead of

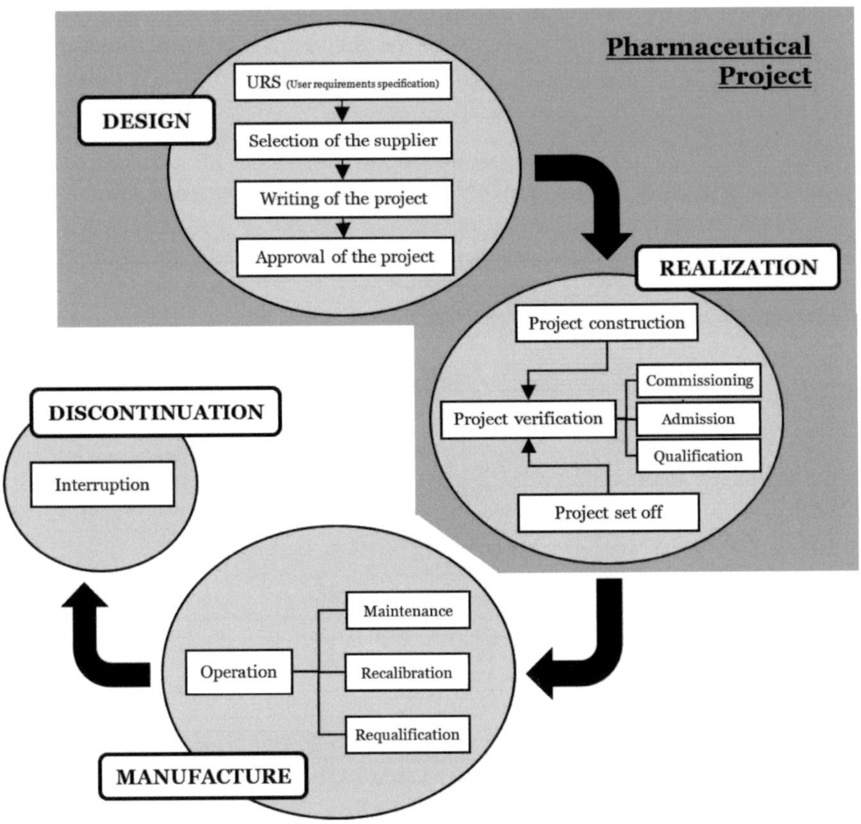

Figure 2.
The pharmaceutical project within the project life cycle.

"manufacture," we can talk of "routine production," where quality is the result of the quality of the project plus all the measures applied to build quality into the products and to monitor that this is correctly realized. The last phase, discontinuation, is simply mentioned to remind us that when facilities have to stop production to be closed, this has to be done in an organized way (i.e., market cannot be left undersupplied by a unilateral decision; the environment has to be protected by recycling or disposing of materials in an ecological way, etc.).

6. Risk management

Figure 3 summarizes the basic definitions related with risk management, as described in guideline ICH Q9 [3]. In any activity, there are hazards, which are significant because they can turn into harms. It is evident that hazards matter because there is a chance that they materialize into harms, and it is obvious that the importance of the harm determines how much attention they deserve. Thus, it is easy to understand the definition of the risk.

There is no human activity free from hazards (and, unfortunately, from harms) and, consequently, risk is always extant. This is why we came to speak of risk management. As we always face risk, it is meaningful to understand it and try to diminish it. When we remove a dish from an (hot) oven, there is a hazard: we can burn our fingers. To diminish the risk, we use an oven glove. By donning a glove, we diminish the probability of the harm and, thus, the risk. And what about the severity of the harm? In general, it is considered that we cannot act against it, as the severity of a harm is an attribute of it. Anyway, we might also think that the simple fact of wearing a glove would diminish the severity of the burn. Yes, in the interpretation of risk, there is always some amount of personal understanding, but this is not very relevant if we come to appreciate the situation well and we apply the same criteria over time. Just to finish with these considerations, we should keep in mind that the exact assessment of risk is second to the accurate understanding of the hazard. This is why in many cases we need not determine risk but just assess the hazards (Which hazards exist? Which are their causes? Are they likely? After all, should we worry? Should we take any measure? etc.).

Figure 4 summarizes the different levels of risk or of hazard assessment (to express it in a different way). As we can see, the objective is always reducing the level of risk as much as possible. This reduction, however, should be consistent with the efforts, which we apply for attaining this effect. Sometimes, a high risk might be accepted if there is no better alternative.

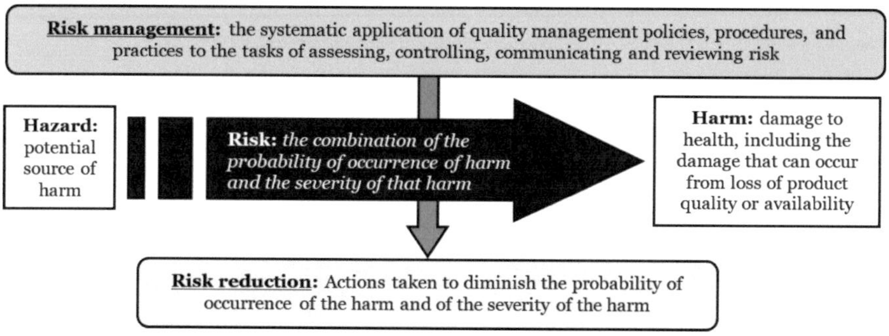

Figure 3.
Main definitions concerning risk management.

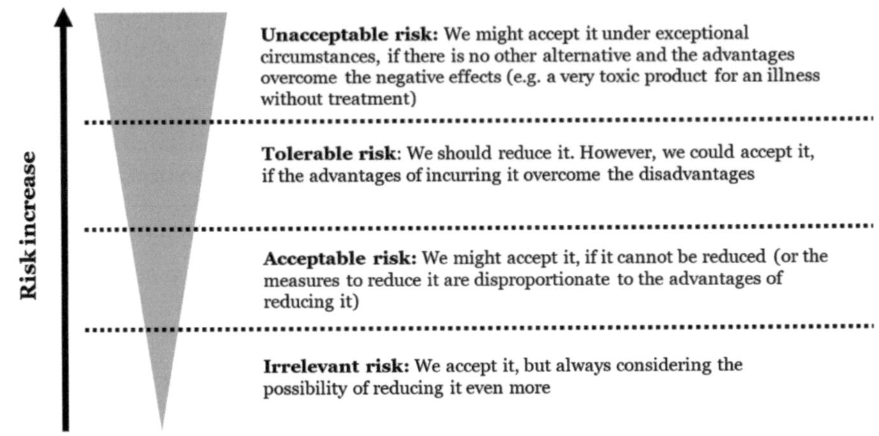

Figure 4.
Risk.

There are several tools, which we can use for risk analysis [3]. They can be used and combined according to particular needs. In fact, the application of a standard method is necessary when high levels of formality are required (e.g., in comparative studies or in scientific papers), but for the routine risk assessment within a company, it is possible to be less formal and adapt the tools to better fit our needs. Tools just organize information. They do not improve our information (if our raw data are poor or inaccurate, the application of the best of tools will not mend them). Thus, the quality of a risk analysis depends mainly on how worth our information is and on the knowledge and experience of the person who performs the study.

Figure 5 lists the most common methodologies used in risk analysis. Among the specific ones, the most popular is, without any doubt, FMECA, an excellent tool when the process under study is well known. Then, it is possible to evaluate risk and use its value as an indicator for process improvement. Anyway, in most cases, when the amount of knowledge is more modest, it is better to start with PHA. HACCP is a very good tool for the control of processes; in fact, the WHO recommends it for

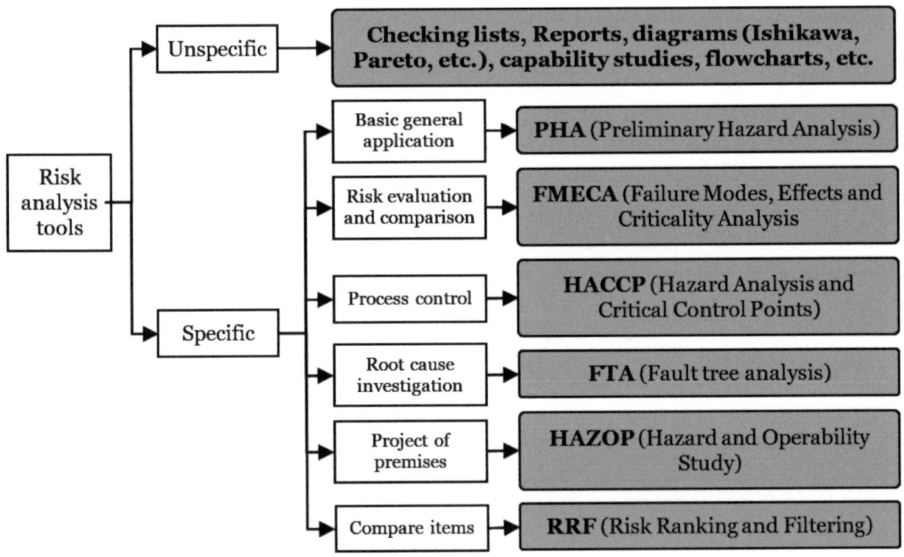

Figure 5.
Risk analysis tools.

the control of pharmaceutical manufacturing processes [6]. It is not necessary to add that HACCP requires a deep knowledge on the process. FTA is useful to identify the root causes of a problem. HAZOP can help to identify possible problems related to equipment and operation involved in a process. RRF is the choice tool for the comparison of items composed of different elements.

Next to the specific methodologies, we can talk of unspecific ones. Properly speaking they are not risk analysis tools but provide information, which is critical to perform an analysis. Here, we should emphasize flowcharts, which are a key element to start a risk analysis of a process.

For the risk management of pharmaceutical projects, we select, besides flowcharts, RRF and PHA. This is why, we consider useful to provide some guidance on them.

Flowcharts are very useful to get a clear idea of a process and to perform a hazard assessment. In fact, the layout of a pharmaceutical unit is the translation of the flowchart steps into premises. The example of flowchart, which we present here in **Figure 6**, covers in a simplified way the operational stages of a pharmaceutical unit. It is evident that in practice there can be different types of processes.

RRF (**Figure 7**) is a tool conceived for the comparison of different items, possessing different types of hazards and risk levels, by reducing them to a common denominator. The application of RRF starts by identifying the items to be compared and identifying their components and subcomponents. Then, the risk factors are determined and their integration allows for the establishment of the global risk of the item. As shown in **Figure 7**, all Rf are given the same weight and, thus, the load of each component depends more on the number of Rf considered than on the importance of the component itself. This can be corrected, for example, by giving a higher classification of risk to single Rf or to all the Rf of a given component.

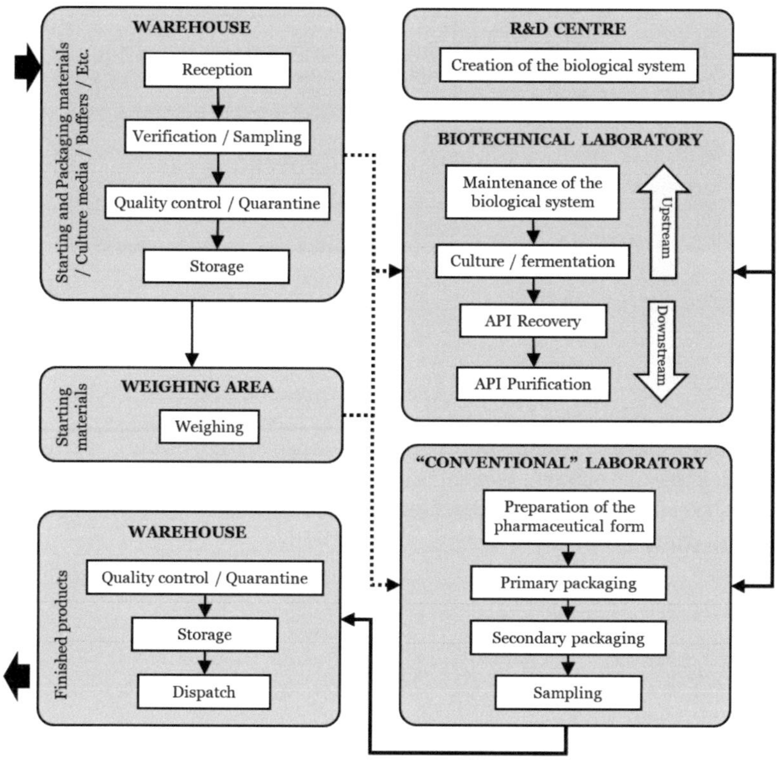

Figure 6.
Flowchart summarizing the steps of pharmaceutical manufacture.

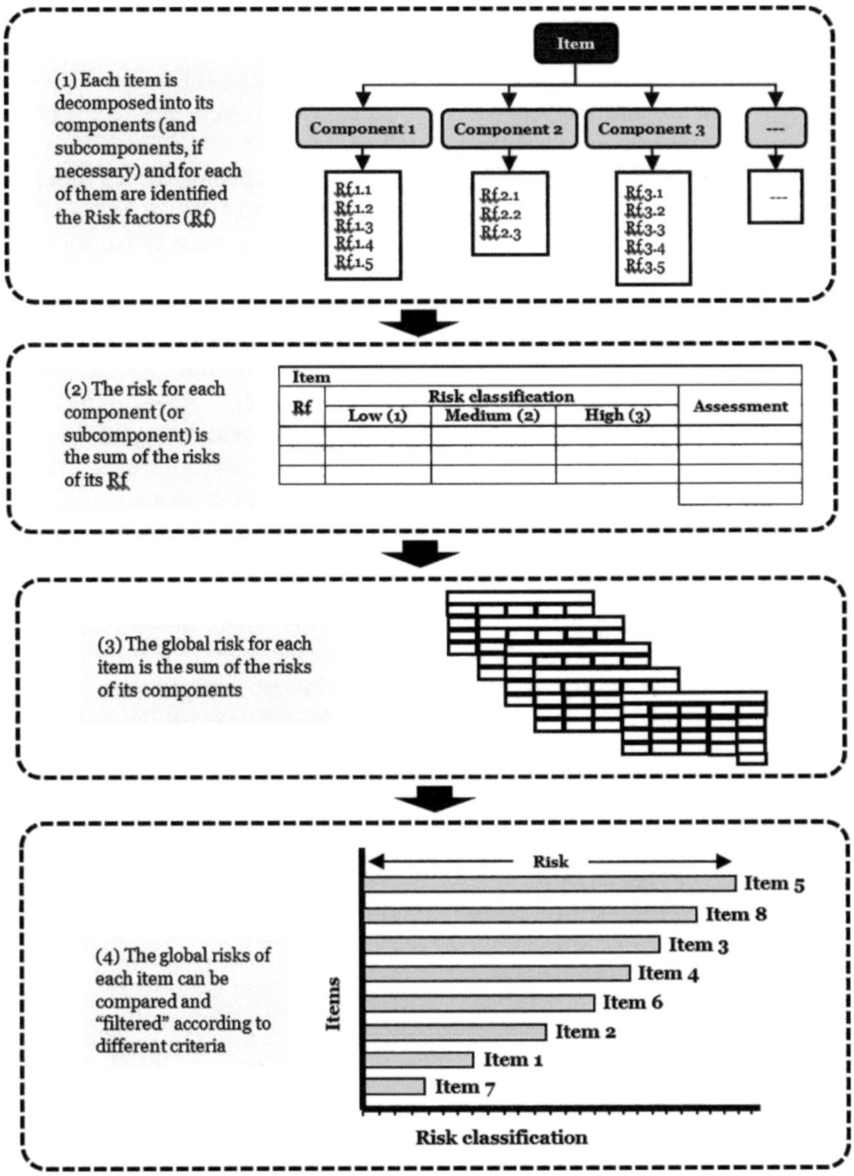

Figure 7.
Summary of steps for the realization of RRF.

PHA is very practical for the analysis of situations where there is still limited information. PHA uses charts like the one shown in **Table 1** (although, often, the column describing the effect can be omitted, because it does not provide any useful information).

Stage/ subject	Possible hazard	Hazard cause	Possible effect (harm)	Is hazard significant?	Control measures
				□ Yes/□ No	
				□ Yes/□ No	

Table 1.
Example of chart used to develop PHA.

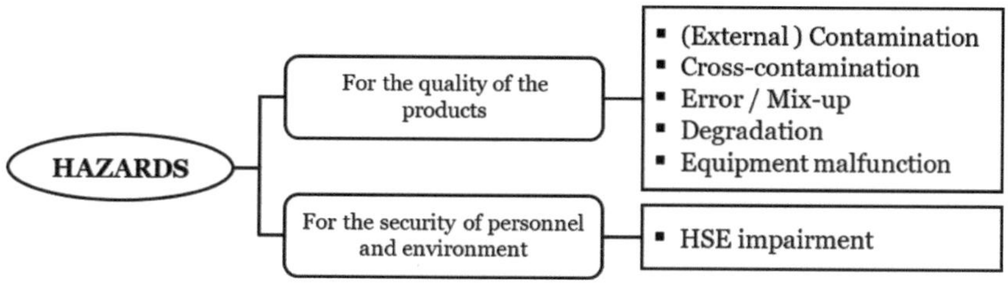

Figure 8.
Summary of hazards to be considered in PHA charts.

Practice shows that, while performing risk analysis, one of the points that creates more confusion is the clear distinction among hazard, cause of the hazard, and effect of the hazard (harm). To simplify this matter, we propose the following approach.

We establish six types of hazards as the base of our analysis (**Figure 8**), then we apply them to the item that we study and we determine if this hazard might exist. If it really exists, then it is easier to establish possible causes and their derived effects and to propose control measures.

7. Pharmaceutical project management

7.1 Supplier selection

It is not necessary to insist on the fact that the selection of the partner with whom a laboratory will realize a project is very important. Following, we provide a simplified example of application of RRF.

Component (1) "Quality"

Rf 1.1 = quality system; Rf 1.2 = training program; Rf 1.3 = realization of commissioning; Rf 1.4 = support for qualification; etc.

Component (2) "Reliability"

Rf 2.1 = experience in projects; Rf 2.2 = amount and completeness of documentation; Rf 2.3 = fulfillment of scheduled requirements; etc.

Component (3) "Accessibility"

Rf 3.1 = number of people in the company; Rf 3.2 = distance of the nearest point of service of the company to the project site; Rf 3.3 = knowledge of the language of the site; Rf 3.4 = after sale service; etc.

Component (4) "Budget"

Rf 4.1 = price; Rf 4.2 = payment conditions; Rf 4.3 = bonuses; etc.

After determining the risk factors for each component, they are used to calculate its comprehensive risk.

Then, the global risk of each item is evaluated adding the risks of their components.

Finally, the items are ranked according to their respective risk. Thus, they can be compared. They can also be "filtered," that is, selected (**Figure 6**).

7.2 Translation of the URS into a plan

As shown in **Figure 9**, a pharmaceutical plan is the practical translation of the requirements set up in the URS.

The conditions penned in the URS depend on the particular wishes of each laboratory, whereas the plan to be developed and constructed by the engineering is

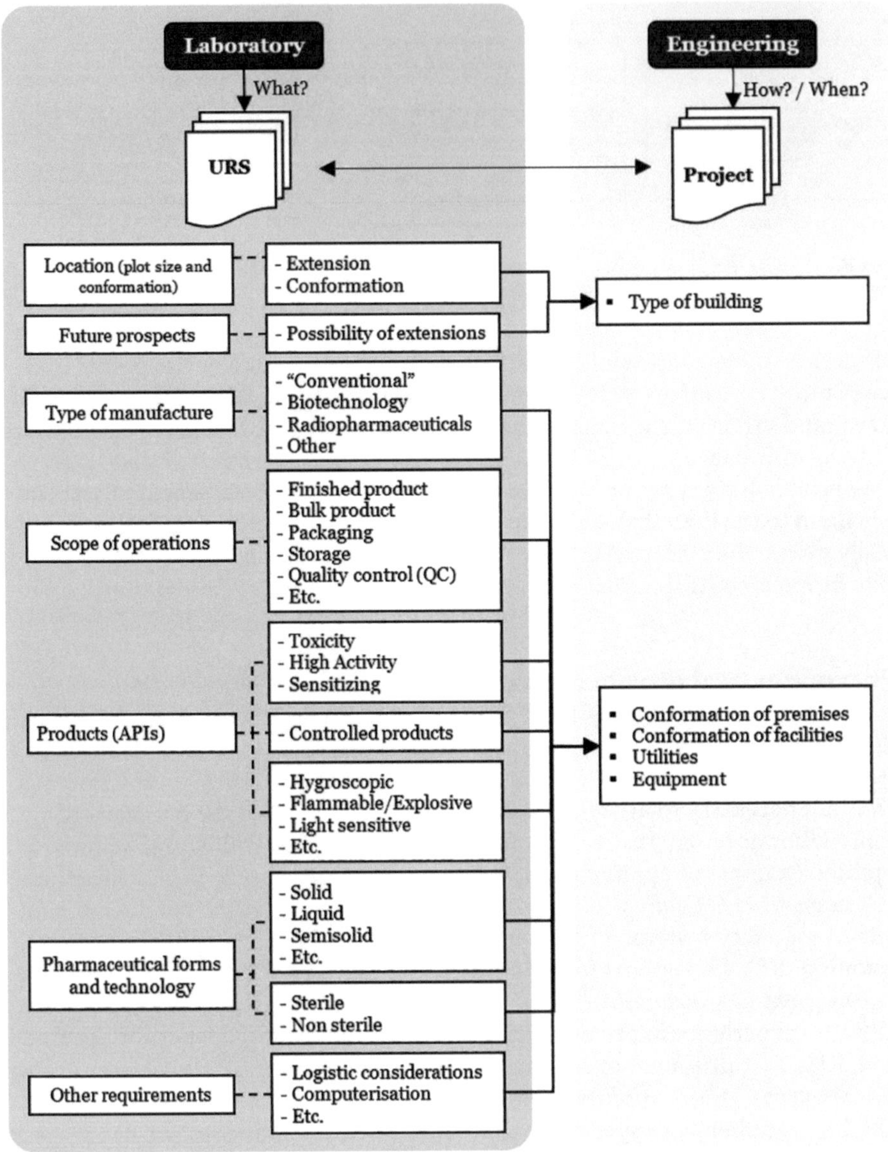

Figure 9.
The pharmaceutical project: URS and their translation.

encoded, by GMP [7–10] and good engineering practice (GEP). GEP is not codified as such. It is understood as the generally admitted good approach.

Let us analyze, using a risk management approach, the requirements that should meet a project. We use an adapted form of PHA.

7.2.1 Hazard #1: (external) contamination

In **Table 2** are described the main causes of (external) contamination and the control measures to keep them at bay.

7.2.2 Hazard #2: cross-contamination

In **Table 3** are described the main causes of cross-contamination and the control measures to hold them at bay.

Cause	Control measures
Inadequate siting of the building	Pharmaceutical units should not be located in contaminated areas
Insufficient tightness of the premises	Ensure isolation of premises from the outside (sealed panels, floors, and windows)
	Access to premises by airlocks/changing rooms
	Protection of premises from the entrance of insects, birds, and animals
	Anticipate the placement of traps and baits
Inadequate separations	Rest rooms and refectories should be separated from areas of production and quality control (QC) laboratory
	Toilets should not have direct communication with the areas of production or storage
	Maintenance workshops should be separated from the production areas
	Parts and tools used for production should be kept in separate rooms or in lockers
	Animal houses should be separated from the other areas
Dirty incoming materials	Receiving and dispatch bays should be designed and prepared to allow the cleaning of the incoming containers
Access through drains	Drains should be designed and built to prevent backflow
Inadequate air-handling	Ventilation air should be HEPA-filtered
	Animal houses should have separate air-handling systems
	Premises should have overpressure (see the exceptions in Sections 7.3 and 7.4) to prevent the entrance of unfiltered air from outside
Defective parts of equipment	Parts of equipment coming into contact with materials and product should not affect them, neither be affected by them

Table 2.
(External) contamination.

7.2.3 Hazard #3: error/mix-up

Table 4 evaluates the main causes of error/mix-up and provides control measures. Although error/mix-up appears because of the inefficiency of personnel (and this means that their prevention is based on training), an improved design of the premises diminishes their probability. Adequate flows and sufficient working space should always be considered in a project.

7.2.4 Hazard #4: degradation

Materials and products are damaged when exposed to inadequate conditions. **Table 5** summarizes control measures.

7.2.5 Hazard #5: equipment malfunction

During routine production, equipment is submitted to a maintenance plan to avoid malfunction. During the project, however, equipment has to be well sited and mounted, as summarized in **Table 6**.

7.2.6 Hazard #4: health, safety, and environment

The denomination health, safety, and environment (HSE) covers all the aspects that can affect the health and security of personnel and the environment. **Table 7** describes actions to control them.

Cause	Control measures
Inadequate facility design	Premises should be designed to allow for adequate cleaning and sanitization
	Premises should be designed to avoid the build-up of dust and dirt
	Repair and maintenance operations should not affect the quality of the products (e.g., performed from outside the clean rooms). Thus, there should be technical areas for equipment and a technical space over the working areas for ducting, lightning, etc.
	The layout of the premises should allow the production to take place in areas connected in the logical sequence of the required levels of cleanliness (e.g., of classification)
	Materials and products should only be exposed to the environment in clean rooms possessing sanitary design
	Packaging areas should be designed and laid out to avoid cross-contamination
	The QC laboratory should be designed to suit the operations
Inadequate tightness of the premises	Clean rooms should be tight to ensure adequate isolation (sealed panels, floors, and windows)
Inadequate separations	Access to classified areas by specific airlocks/changing rooms
	There should be an area for the sampling of starting materials
	There should be an area for the weighing of starting materials
	Risk of cross-contamination by highly active, toxic substances or biological agents should be controlled. See Section 7.3
	The QC laboratory should be separated from the production areas
	The QC laboratory areas where microbiological, biological, and radioisotope tests are performed should be separated from each other
Inadequate cleaning	There should be an adequate cleaning area with adequate separations for equipment to be cleaned, cleaning area, drying area, and storage of clean equipment
Inadequate air-handling	When operations are likely to generate dust (e.g., sampling, weighing, mixing, etc.), there should be measures to control it
	Starting materials should be weighed in a special area provided with unidirectional flow and exhausting
	Materials and products should only be exposed to the environment in clean rooms possessing appropriate air-handling
	Recycled air should be HEPA-filtered
	Pressure differentials should control the flows of air among clean rooms
	The QC laboratory should possess an adequate separate air-handling system
	The QC laboratory areas where microbiological, biological, and radioisotope tests are performed should possess separate air-handling systems
Inadequate equipment	Equipment should be installed to avoid contamination
	Washing, cleaning, and drying equipment should not be source of contamination
	Whenever possible, closed equipment should be preferred

Table 3.
Cross-contamination.

Cause	Control measures
Inadequate facility design	Layout and design must aim to minimize the possibility of errors
	Premises should be designed to be able to follow the logical flows of personnel and materials
	The storage areas should allow for the orderly and sure storage of the different sorts of materials and products ("traditional warehouse") or possess a computerized management system providing the same level of security ("chaotic warehouse")
	The layout of the premises should allow the production to take place in areas connected in the logical sequence of the operations
	The areas should permit the orderly and local positioning of equipment and materials
	In the production areas, there should be in-process storage rooms permitting to keep materials and equipment in an orderly way
	Packaging areas should be designed and laid out to avoid mix-ups
	The QC laboratory should have sufficient space
Inadequate separation	The storage areas should ensure the segregation for items under quarantine, returned, rejected, and recalled. This separation can be assured by separated and closed areas or by a computerized management system providing the same level of segregation
	The receiving and dispatch bays of the warehouse should be separated
	Printed materials should be stored in a separate area with restricted access
Free access	Personnel access to critical areas should be controlled and restricted
Inadequate marking	Fixed pipework should be labeled to indicate contents and direction of flow (if this is necessary)
	Rooms and equipment should be adequately identified
Inadequate illumination	Working areas should be well lit
Inadequate placement of equipment	Equipment should be installed to avoid error and mix-up

Table 4.
Error/mix-up.

Cause	Control measures
Inadequate facility design	Receiving and dispatch bays should protect the products from the weather
Inadequate conditions	Acquire adequate information on the product requirements
	Temperature and, if necessary, humidity in the production areas should be adequate and monitored
	Electrical supply, lighting, temperature, humidity, and ventilation should not adversely affect the products
	Storage areas should provide adequate conditions of temperature (and if necessary humidity) for the materials and products

Table 5.
Degradation.

Cause	Control measures
Inadequate installation	Perform commissioning/qualification
	Equipment must be located to suit the operations
Inadequate separation	Consider in the QC laboratory separate rooms for instrumentation, in accordance with their particular requirements

Table 6.
Equipment malfunction.

7.3 Management of toxic substances

Sometimes commercial/logistic aspects determine from the outset the characteristics of a pharmaceutical plant: typically, multiproduct facilities (when APIs are not particularly active and general GMP precautions suffice for ensuring that significant

Cause	Control measures
Inadequate installation	Perform commissioning/qualification
	Equipment must be located to suit the operations
Dangerous products	Highly active, toxic substances or biological agents should be controlled (see Section 7.3)
	There should be special storage areas, safe and secure, for: highly active, radioactive, narcotics, abuse, explosive, flammable, etc.
Dangerous biological agents	Biological agents should be handled adequately (see Section 7.4)
	There should be adequate equipment (freezers, refrigerators, etc.) for the maintenance of the biological agents
Inaccessibility/unhandiness	Premises should be designed to suit the operations to be carried out
	Changing rooms and toilets should be easily accessible and adapted to the number of users
	Pipework, light fittings, ventilation points, and other services should be designed and placed to avoid the creation of recesses difficult to clean. And, as far as possible, they should be accessible from technical areas
	Open channels should be avoided where possible, but if they are necessary, they should be shallow to facilitate cleaning
	The QC laboratory should be designed to suit the operations
	In the QC laboratory, there should be adequate space for the storage of samples, standards, solvents, reagents, and records
	Current drawings of critical equipment and utilities should be maintained
Uncontrolled solid waste	There should be adequate places for the storage of solid waste
	There should be a place for the classification of solid waste
Uncontrolled effluents	There should be a place for the treatment of liquid effluents (decontamination) prior to its release or transportation to a handling center
	Dangerous gaseous effluents should be either filtered or incinerated

Table 7.
HSE.

cross-contamination will be excluded) or, less frequently, dedicated facilities, when logistic reasons recommend limiting the number of products in order to increase output. A quite different situation may arise when APIs can be deemed "toxic." This term is used in a practical way to provide a general denomination for substances that possess high activity or potency (e.g., tiny amounts are needed to produce a pharmacological effect) or harmful effects (e.g., sensitization, genotoxicity, etc.).

When designing premises where toxic APIs will be handled, there has always existed a key question: is it correct if they are multiproduct or we should opt for dedicated ones? The response to this question cannot be general, because substances are diverse. The traditional approach was considering, roughly speaking, three cases. Firstly, we had the APIs, which could be deemed nontoxic and which, as we said before, could be produced in multiproduct facilities. Secondly, we had the APIs possessing high activity (e.g., hormones, cytostatics, certain antibiotics, etc.), which required dedicated facilities. And finally, we had two cases, which required strict segregation. This last group included live microorganisms and products possessing sensitizing or toxic effects (their action, properly speaking, cannot be quantified and should be considered as "on/off"), such as beta-lactam antibiotics [11].

In order to better clarify this group, it was proposed a scientific approach, which is summarized in **Figure 10**. The flowchart combines an EMA guideline on this matter [12] (upper part with a darker shade) with logistic criteria (lower part).

When it is spoken of single product or multiproduct facilities, their meaning appears evident, but what about dedicated facilities? This implies separation, but what kind of separation. Only risk management can provide an adequate answer to this question. **Figure 11** describes the rationale of this approach. The rectangle represents the risk level (on the left lower part, the lighter color indicates low risk, whereas, on the right upper part, the darker color shows high risk). Thus, multiproduct facilities are adequate when risk is low, and segregated facilities are necessary when risk is high. Then, in the middle, where risk can be deemed medium, it is possible to think of intermediate solutions (e.g., instead of separation of facilities, separation of products) or campaign production (e.g., separation is not physical but temporal). The severity of the hazard depends on the API, whereas the probability of occurrence is related to the way of manufacturing. In other words, once the "toxicity" of the API is known, the practical level of risk will depend on the production techniques used for the manufacture of the products. Dedicated facilities mean more expensive projects, but at the same time lower risk of cross-contamination and, for instance, easier and surer cleaning validations [13].

7.4 Management of biological systems

Biotechnology implies the use of biological systems. Under this term, we design both cells and microorganisms. Most of them (in principle, the cells, and many microorganisms) do not pose any particular thread to personnel. In fact, the contrary is true. They are labile and very susceptible to contamination and require strict measures of control to keep them viable. There are, however, microorganisms, which suppose a thread for the personnel if they infect them during operations. These "biological agents" are internationally classified into four groups (**Table 8**) in function of the level of biosecurity (BSL) or protection level (PL) that they require [14, 15].

The project of a laboratory handling biological agents has to take into account its two types of requirements: on the one hand, those regarding in general a pharmaceutical laboratory and on the other, the particular necessities of facilities containing live microorganisms. We have already considered the former; thus, here we will exclusively discuss the latter.

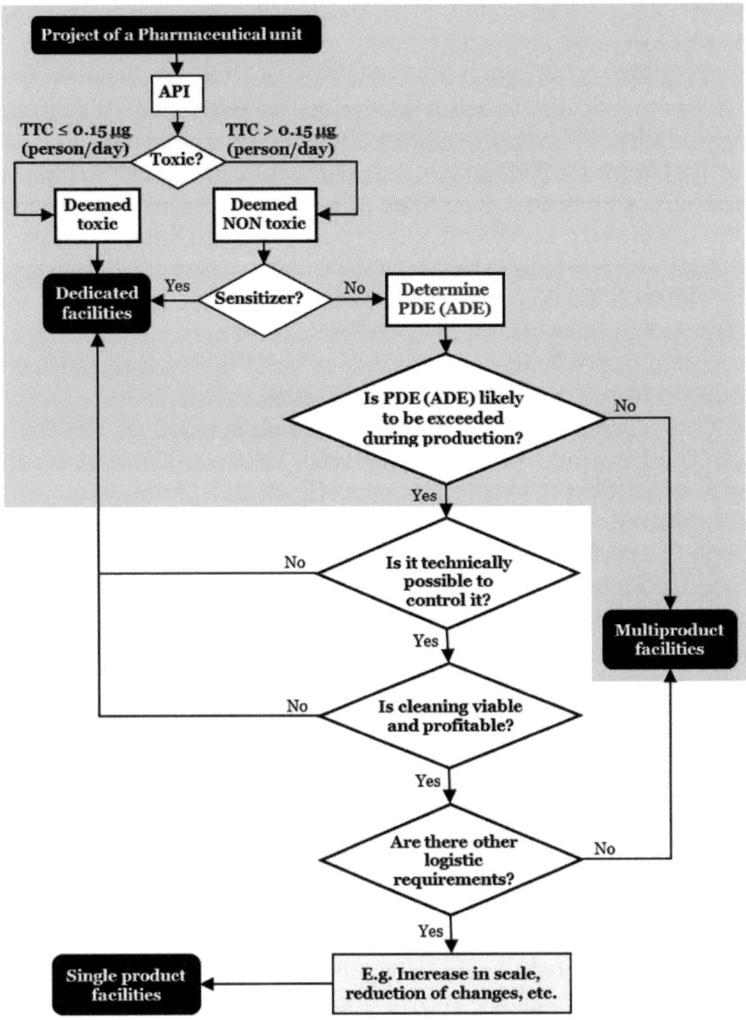

Figure 10.
Decision tree for facilities.

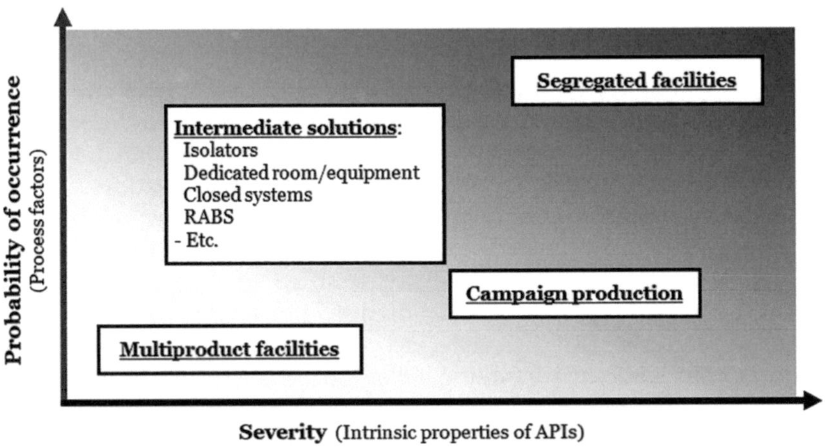

Figure 11.
Risk analysis rationale for defining the type of facilities.

Characteristic	**BSL/PL**			
	1	**2**	**3**	**4**
Personal risk	Minimal	Low	High	Very high
Epidemic risk	No	No	Low	High
Therapy/prevention	Unnecessary	Available	Available	Unavailable
Manipulation	Workbench	Open hood	Biological safety cabin (BSC)	Isolator or BSC + personal protective equipment (PPE)

Table 8.
The four groups of biological agents.

In a laboratory where biological agents are cultured, these microorganisms suppose the "hazard"; the "cause" is the inadequacy of the measures taken to ensure that these agents will remain contained; and the "effect" is the infection of people by released agents. Thus, our risk analysis will focus on the control measures (**Table 9**).

Element	**Control measures**
Premises	Area segregation
	Ensure contention
	Ensure possibility of disinfection
	Operations can be seen from the outside
	Easy communication with the exterior (e.g., interphone)
	Ensure security in case of emergency (e.g., earthquake, flood, fire, etc.)
Air-handling system	HEPA-filtered/sterilized exhausted air
	Ensure depression ($\Delta P-$)
	No recycling of air
Ingress	Restricted access
	Airlocks with interlocked doors
	Pass-boxes with disinfection systems
	Sterilizers provided with double doors
	Separated changing rooms for entry and exit
Equipment/clean rooms	Ensure operation of critical equipment in case of power supply cut
	Provide separation between agents and operators
	Provide wash basins with hand-free taps
	Vacuum tubes protected with HEPA filters or disinfectant traps
Waste	Inactivation of "biowaste"
	Inactivation of effluents
Labeling	The international sign of biological hazard should be affixed at the entrance of the laboratory and at critical rooms and critical equipment (e.g., incubators, freezers, etc.).

Table 9.
Control measures for biological hazard in the laboratory.

In **Table 9**, we mention a series of elements that should be taken into account when designing a laboratory manipulating biological agents. The characteristics of these elements depend on the BSL/PL. This means that they might be unnecessary for level 1, just recommended for an intermediate level and required for a high level. It has to be studied case by case, using a risk management approach, which should analyze the level of risk and the level of protection provided for the systems in place.

8. Conclusions

Society requires medicines possessing the purported quality, and this is only possible if we design, manufacture, monitor, and control them adequately. These undertakings, however, involve appropriately devised, built, and qualified premises and facilities, which are the outcome of a project. A pharmaceutical project is a very complex subject because it involves at least two parts, with different visions and experience, and because it implies scores of suppositions and projections into the future. The risk of straying away from the expected roadmap is quite real. Thus, risk management becomes necessary. In a project, the amount of information is limited and thus, while using risk analysis tools, it is essential to bear in mind that what really matters is to understand which problems are at stake and to get a clear picture of their respective relevance. Once this is achieved, it is easier to provide adequate solutions. Risk is controllable, if we know and understand it!

A frequent problem, when trying to evaluate risk, is that information is so varied and multiform that it becomes difficult to ascertain, which are the hazards and distinguish them from, say, causes and effects, let alone the logical organization of the information. To overcome these difficulties, we propose two tiers of solutions: on one side, to use simple risk analysis tools focusing more on hazard and control measures than on risk quantification and on the other, to use defined simple quality hazards as the point of departure for the analysis.

Conflict of interest

None.

Author details

Jordi Botet
GMP Consultant, Barcelona, Spain

*Address all correspondence to: jbotetfregola@gmail.com

References

[1] ICH (International Council for Harmonisation of Technical Requirements for Pharmaceuticals for Human Use). Pharmaceutical Quality System. Harmonised Guideline Q10. Geneva, Switzerland: ICH; 2008

[2] ICH. Pharmaceutical Development. Harmonised Guideline Q8(R2). Geneva, Switzerland: ICH; 2009

[3] ICH. Quality Risk Management. Harmonised Guideline Q9. Geneva, Switzerland: ICH; 2005

[4] ICH. Development and Manufacture of Drug Substances (Chemical Entities and Biotechnological/Biological Entities). Harmonised Guideline Q11. Geneva, Switzerland: ICH; 2012

[5] ICH. Technical and Regulatory Considerations for Pharmaceutical Product Lifecycle Management. Harmonised Guideline Q12, Draft. Geneva, Switzerland: ICH; 2017

[6] WHO (World Health Organisation). Guidelines on quality risk management. In: WHO Expert Committee on Specifications for Pharmaceutical Preparations. Forty-Seventh Report (WHO Technical Report Series, No. 981). Annex 2. Geneva, Switzerland: WHO; 2013

[7] WHO. Good manufacturing practices for pharmaceutical products: Main principles. In: WHO Expert Committee on Specifications for Pharmaceutical Preparations. Forty-Eighth Report (WHO Technical Report Series, No. 986). Annex 2. Geneva, Switzerland: WHO; 2014

[8] European Commission. Good Manufacturing Practices. Medicinal Products for Human and Veterinary Use. The Rules Governing Medicinal Products in the European Union. Vol. 4. Brussels. http: //ec.europa.eu/health/documents/eudralex/vol-4_en

[9] US National Archives & Records Administration: Federal Register. Code of Federal Regulations (CFR). Title 21 (Food & drugs). Parts 210 through 226. https: //www.govinfo.gov

[10] WHO. Good manufacturing practices for biological products. In: WHO Expert Committee on Specifications for Pharmaceutical Preparations. Thirty-Third Report (WHO Technical Report Series, No. 996). Annex 3. Geneva, Switzerland: WHO; 2016

[11] U.S. Department of Health and Human Services. Non-Penicillin Beta-Lactam Drugs: A CGMP Framework for Preventing Cross-Contamination. Guidance for Industry. Current Good Manufacturing Practices (CGMPs). Rockville, MD, USA: Food and Drug Administration (FDA). Center for Drug Evaluation and Research (CDER); 2013

[12] EMA (European Medicines Agency). CHMP (Committee for Medicinal Products for Human Use), CVMP (Committee for Medicinal Products for Veterinary Use). Guideline on Setting Health Based Exposure Limits for Use in Risk Identification in the Manufacture of Different Medicinal Products in Shared Facilities. EMA/CHMP/CVMP/SWP/169430/2012. London: EMA; 2014

[13] WHO. Good manufacturing practices for pharmaceutical products containing hazardous substances. In: WHO Expert Committee on Specifications for Pharmaceutical Preparations. Forty-Fourth Report (WHO Technical Report Series, No. 957). Annex 3. Geneva, Switzerland: WHO; 2010

[14] US Department of Health and Human Services. Biosafety in Microbiology

and Biomedical Laboratories. HHS Publication No. (CDC) 21-1112. 5th edition revised ed. Atlanta, Georgia, USA: Public Health Service Centers for Disease Control and Prevention. National Institutes of Health (CDC); 2009

[15] WHO. Laboratory Biosafety Manual. 3rd ed. Geneva, Switzerland: WHO; 2004

Section 4

Risk, the Future Paradigm

Chapter 10

Paradigms of Risk, Hazards and Danger

Marek Różycki

Abstract

In his 1962 work, The Structure of Scientific Revolutions, Thomas Kuhn defined paradigm as a set of concepts constituting the foundations of a field of science. He presents revolutions as shifts in the existing paradigms and the phrase, paradigm shift, has since entered the language of science and business. Risk is a concern in both fields and this chapter considers the paradigms of risk, and whether they require a shift. Although we avoid negative experiences, often interpreted as resulting from hazards, no common risk management methodology exists. This statement may strike as untrue: after all, safety is a vast field; we analyse hazards and manage risk. Yet is it not a delusion, and is risk management not an attempt to charm reality? Don't hazards, risk and danger depend on our perception? Perhaps risk can be viewed through the lens of quantum mechanics, existing in a limbo of potential until our actions and interpretations force events and circumstances to assume a danger state. If so, would managing this potential prior to the wave function collapse-inducing observation make any sense? In this chapter we will use the theory of inertia to attempt an answer to the question: is risk management possible?

Keywords: paradigms, risk, hazards, danger, theory of inertia

1. Introduction

We usurp the right to manage future events: we decide where to meet friends for dinner, to which schools to send our children and where to go on holiday. We choose jobs, hobbies and spouses. We take decisions and change our reality: we feel the masters of our universe, except when a spanner, carelessly thrown, wrecks the carefully planned works. On the way to the dinner, our bus gets stuck in a jam. We miss the holiday flight because we have misplaced the car keys, or worse, a fire breaks out in our DIY shed and spreads to the house, and the roof caves in over our spouse; on the way to a job interview, a tree topples on top of us, or we fall into a manhole.

We try to prevent such misfortunes and proudly call our attempts for risk management. We manage risks in the blind faith that there is a fate which we can outsmart. Yet there is only so much we can control, and it may be that, preoccupied with the robustness of the roof, we do not notice that a spanner has been misplaced on a high shelf. On our way out to dinner, we slam the door to the DIY shed, and the spanner, balancing precariously until now, finally changes state and falls into a toaster; we get the picture—observe it—after the event and blame fate without noticing that no such thing exists.

2. Paradigms of risk, hazards and danger

2.1 Man proposes, god disposes?

Since the cultural revolution of the Renaissance, we have arrogated power over our world and have come to believe that we can influence future events. From lotteries and stock markets through to diarised work meetings, we always try to increase the certainty of success. We expect to manage the future, based on past events. This belief in our potential to manage our future can be captured as follows:

Pessimistic tendencies are thought to be related to our survival instinct—the better you can foresee a misfortune, the better your chances—optimism can lead to ignoring past experiences. According to this theory, pessimists should live longer and more stable lives.

At the level of the individual, medicine does not corroborate this theory, and the adverse psychological effects of pessimism on the health might be at play. However at the collective level, pessimism in the form of careful assessment of reality, hazards, the future and the associated risks, appears to be the dominant modus operandi.

We may form the following observations based on **Figure 1**:

1. Possible events are not unlimited, and the number of possible events depends on the correlations between the relevant factors, with varying threshold limit values to each combination. It is also possible to define threshold properties of events, based on their combinations, although it is not always possible to identify individual events.

2. Human experience is based on concrete past events, recorded in the individual and the collective memory. Describing history as it does, this data set also informs us about the possible futures: the greater the set, the more possibilities for consideration. However, the set is never complete—giving rise to the black swan phenomenon, as well as Sod's/Murphy's law at lesser scales. This appears to be congruous with the wider laws of physics[1].

3. Identification of future scenarios is a combination of empirical and creative endeavours, and it used to be assumed that computational feasibility has little influence over it. However, if we consider historical data and threshold properties of possible future events, our calculations will give us the edge in preparing for identified possible outcomes. The 2016 Bastille Day lorry attack in Nice, France, may serve as an example of the cognitive deficit in analysing future events: security services were well prepared for potential terrorist attacks before and during the fireworks display, but not after the display; that is when

[1] According to the second law of thermodynamics, every system and process will suffer energy dispersion tending towards disorder, or entropic equilibrium, simply because disorder is more probable. If we connect two vessels, one containing oxygen and the other containing nitrogen, it is improbable that after connecting, both gases will remain in their original vessels. Such an event is possible, but it is more probable that the atoms of both gases will disperse across the two vessels.

the terrorist drove the lorry into the crowd [1, 2]. Any method comprising creative combinations of analysis and intuition could have allowed such a scenario to be imagined and prepared for. Stock market analysis is another good example: the more complex the model, the better the analyses—which still does not rule out error. Share prices, as a collective expression of human actions, are variable and frequently unpredictable. However, if we consider the possibility of both a rise and a fall of share prices, we will survive.

4. Known scenarios (as expressions of the applied computational feasibility) will be appropriate to the degree to which past experiences and possible events are considered.

5. If, at the stage of future event assessment (identifying scenarios), we fail to consider possible events not based on experience, our analysis will be flawed.

It is impossible to avoid the limiting influence of experience; therefore, no risk assessment is fully rational. Managers should regard risk assessment as ancillary and not the foundation of decision-making.

The above statement may be illustrated by considering catastrophes. The list below consists only of incidents caused by ignoring known and predictable events and circumstances (**Table 1**).

The disasters were caused by events and circumstances which were not past experiences but were nevertheless predictable. The disasters themselves may be deemed unpredictable because the events which led to them slipped out of control. It is possible to identify analogous events elsewhere which were rectified because the processes remained under control—for example, in the 1979 incident at the Three Mile Island Nuclear Generating Station in Pennsylvania, USA. The effects of human actions may be deemed unpredictable, even though the scenarios which ensue were predictable. Does only God dispose what man proposes? We can only blame gods or fate if our risk assessment model is limited to analysing known past events. As our computational capabilities grow, we are able to analyse more and more past events and identify scenarios more accurately.

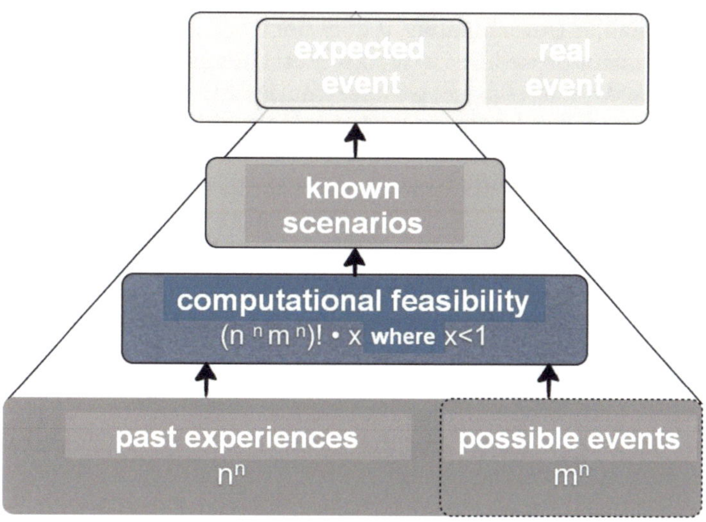

Figure 1.
Determinants of risk management decisions.

Event	Identified cause	Real cause
Courrières mine disaster 10 March 1906	Coal dust explosion 1099 deaths	Use of naked-flame miners' lamps [3]
Bhopal gas tragedy 3 December 1984	Methyl isocyanate leak 15,000 dead, 560,000 nonfatal injuries	Backflow of water into a leaky methyl isocyanate tank [4]
Chernobyl nuclear disaster 26 April 1986	Uncontrolled reactor conditions caused core meltdown 31 direct casualties, 350,000 people resettled, fallout area of 140,000 km^2	Planned safety test delay due to temporary electricity supply cut from another power plant, testing the cooling system negligently [5]
Fukushima Daiichi nuclear disaster 11 March 2011	Tsunami protection measures. Historically waves did not exceed 6.1 m	14-metre wave Flooding of fuel tanks for nuclear reactor cooling generators [6]

Table 1.
Industry incidents.

2.2 Hazard or danger?

The terms danger, hazard and risk are frequently used interchangeably. The first two are especially prone to confusion: in most European languages, they are treated as synonyms.

Risk is also considered a synonym to danger and hazard. A home owner will speak of avoiding the risk of fire or the hazard of fire or the danger of fire. However, the insurer will assess the risk, but not the hazard of fire, using a more precise language. Nevertheless, the distinctions are vague and lead to inadequate risk assessment and management procedures. Let us consider the following example.

In many countries, including the European Union, every job requires to have its risks assessed, and employees responsible for this task use different assessment methods. In all these methods, the activities and settings of the job are described, and the risk is calculated in percentages or degrees; we learn, say, that a given job's risk is low or, elsewhere, that damages are acceptable. What does it mean?

Will the employee in the assessed job not have an accident? Or if they do, will the damage be acceptable? Unfortunately, it does not work like that. I have participated in many post-accident procedures in workplaces. Usually the situation pans out as follows: the accident takes place; we have injuries and damages. In all analysed events, the documented risk assessment deemed the risk low and the damages acceptable. Since the accident happened, what was the goal of the assessment?

Majority of assessment methods use statistical models in which we are required to define the probability of an event taking place and the severity of its effects. Looking back at the determinants of risk management decisions shown in **Figure 1**, our work will be purely theoretical and past-focused. Instead, the probability of describing real events which will take place can be calculated with the help of Bernoulli's principle, as shown in **Figure 2**.

Two probabilities coexist here:

P1—the probability of identifying the scenario which will take place.

P2—the probability of identifying the scenario which has taken place.

The joint probability will be a sum:

P1 + P2.

The probability that an event will occur k times following n attempts, where $k = n$, can be described as follows:

$$Pn(k) = \left(\frac{n}{k}\right) p^k p^{n-k} \tag{1}$$

Therefore, we can assume the following values for the probability of identifying the scenario which will take place.

We assume that every time the probability of success (the answer "yes") equals

$$q = \frac{1}{2}, \tag{2}$$

and the probability of failure equals

$$p = \frac{1}{2} \tag{3}$$

P1(1) = 0.5.

We can assume the following values for the probability of identifying the scenario which will take place.

We assume that every time the probability of success (the answer "yes") equals

$$q = \frac{1}{2}, \tag{4}$$

and the probability of failure equals

$$p = \frac{1}{2} \tag{5}$$

Four variants are possible: where the predicted scenario has happened and the event was a success (1/2) or failure (1/2); and where an unpredicted scenario has happened and the event was a success (1/2) or failure (1/2).

P2(1) = 0.25.

There is a causal relationship between events and scenarios, and consequently we can calculate the risk to equal, at best, 0.5*0.25 = 0.75.

Our margin of error is, therefore, 25%, which necessitates eliminating mistakes.

Hazard is the subjective property of a situation or an object. It is independent of the environment and does not need to interact with it. It is describable. A crocodile in a pond will serve as the perfect example: we know that it is alive, a predator, and that meeting it will not be pleasant or neutral to our wellbeing. We can describe the crocodile's physical features, behaviours and habits (**Figure 3**).

The fact of the crocodile's existence implies nothing. Yes, it is a hazard; we can assert this to be true, and that is it.

As soon as the hazardous object or situation begins to interact with our activity, the hazard becomes a danger (**Figure 4**).

A planned or undertaken activity may, on contact with a hazard, cause danger to arise. The danger can be expressed as the measure of injury or damage. As we can see, it is important to differentiate between hazard and danger: if it cannot be quantified in terms of injury or damage, it is only a hazard.

Going back to job risk assessments, we should focus on describing hazards and pointing out dangers caused by the employee's specific activities. Unfortunately, since these concepts are not easily aligned with our image of reality, they rely on

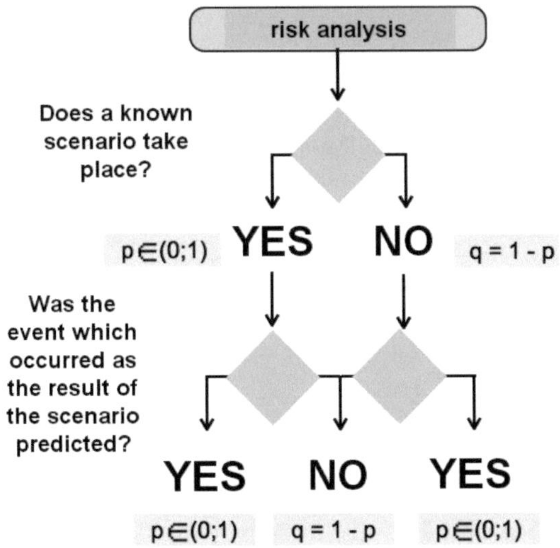

Figure 2.
Probability of identifying the event predicted in risk analysis.

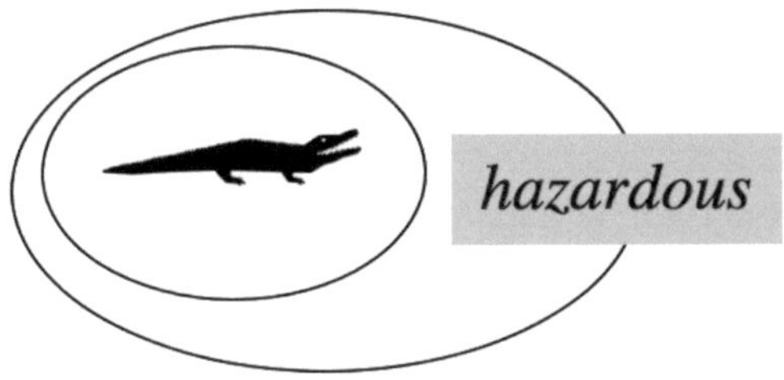

Figure 3.
Visualisation of a hazard: a crocodile in a pond.

Figure 4.
Visualisation of danger: activity reveals the destructive potential of the hazard.

being understood by risk assessors who frequently reach false conclusions regarding our influence over future events.

2.3 Risk

Risk is the potential variability of events [7] or, to put it another way, the influence of uncertainty on goals [8]. Uncertainty can be positive (expected) or negative (unexpected), and goals can concern any human or organisational activity. Certainty is defined as the lack of uncertainty; uncertainty is a mental state characterised by the lack of information which would allow understanding of an event, its results or situational probability. Lack of information is not tantamount to the absence of an event, but results from flawed decision determinants (see **Figure 1**).

Risk, understood as the influence of uncertainty on goals, has the following characteristics:

- It refers to potential events, their results or both.
- It is the combination of an event's probability and its results (including any circumstance changes), and as such it admits the use of statistical tools.

Risk assessment should inform us about the level of uncertainty and should not be considered separate from this factor (**Figure 5**).

Risk should be expressed as a combination of the level of goal attainment certainty (from 0%, no chance of attainment, to 100%, certain attainment) and the level of goal attainment (where 100% means the goal has been fully achieved). When we bring these factors together, we get the risk matrix (**Figure 6**).

The most desirable state is when both the level of goal attainment certainty and the level of goal attainment exceed 50% **although, in the case of negative outcomes in safety or mission-critical domains, the tolerability levels are often set by regulators and may be below 50%**. That's field 1 of the matrix. Any other risk level, captured above in fields 2, 3 and 4, should not be acceptable and should require corrective measures (**Figure 7**).

It is easy to see that the area of acceptable risk is quite small. In the matrix above, it is less than 20%, though the level may vary depending on situation. In majority of cases, however, risk assessment will require corrections to the original assumption. It would be wrong to assess risk merely as a statistic (e.g. as the quotient of effects and probability).

We must therefore assume that in over 80% of cases, risk assessment will require corrections to the original assumptions. It would be wrong to assess risk merely as a statistic (e.g. as the quotient of effects and probability).

Schrödinger's cat, Erwin Schrödinger's famous 1935 thought experiment, will be an excellent example to illustrate the inadequacy of statistics in risk assessment.

Let us assess risk for the following situation:

A cat is penned up in a steel chamber, along with the following device (which must be secured against direct interference by the cat): in a Geiger counter, there is a tiny bit of radioactive substance, so small, that perhaps in the course of the

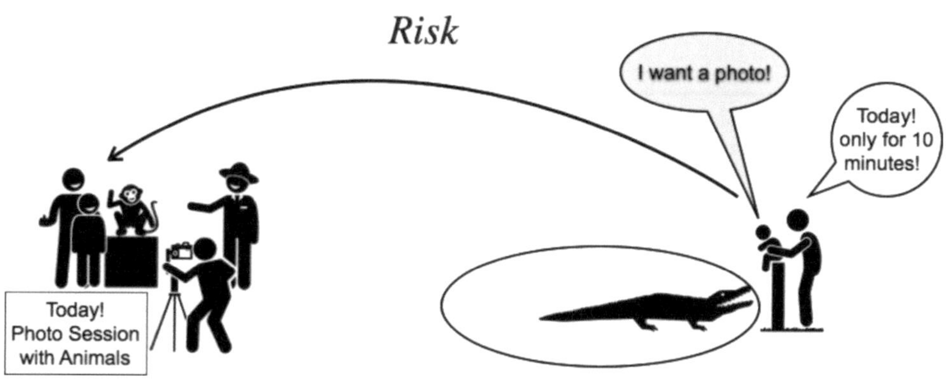

Figure 5.
Visualisation of risk: the influence of the uncertainty of danger (the destructive potential of the hazard) on goals.

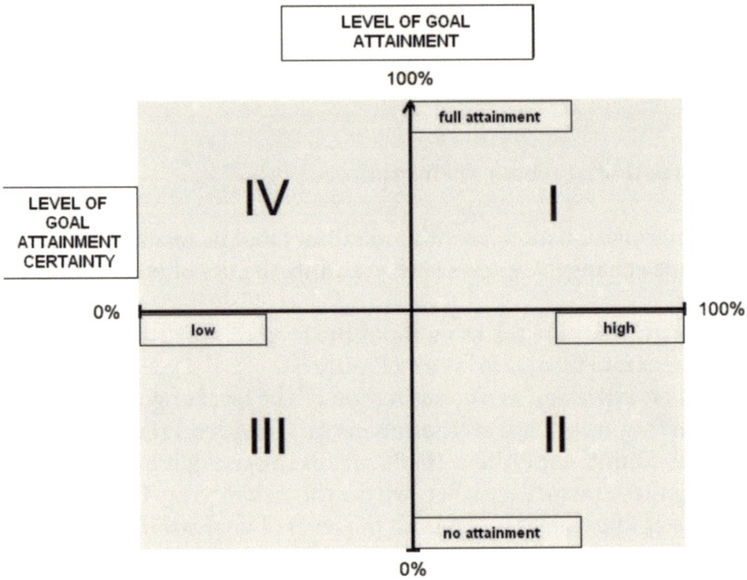

Figure 6.
Risk matrix.

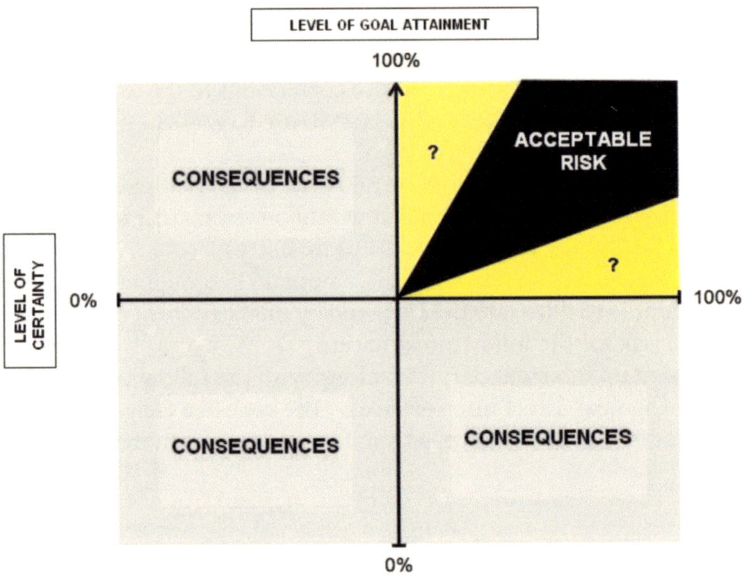

Figure 7.
Risk matrix—practical application.

hour, one of the atoms decays but also, with equal probability, perhaps none; if it happens, the counter tube discharges and through a relay releases a hammer that shatters a small flask of hydrocyanic acid. If one has left this entire system to itself for an hour, one would say that the cat still lives if meanwhile no atom has decayed. The first atomic decay would have poisoned it [9].

2.4 What is the risk that the cat is dead?

The act of observation affects the observed: the accuracy of this famous pronouncement goes beyond quantum mechanics. In psychology, ideas to which we give greater attention tend to become exaggerated and push other ideas out: we think of ourselves as lucky or unlucky, holding a cup that is half-empty or half-full, even though no matter what we think, the reality is one and the same, and only our focus is elsewhere. This focus, as the result of a whole life's experience (see **Figure 1**), may become a major limiting factor for us, the perceivers. Until we open Schrödinger's chamber, the cat may be in any possible state, and so, in the physics of the infinitesimal as much as in risk assessment, it could be both dead and alive. Of course, we can describe the dangers lurking in the chamber, for the cat and for us, and we can also point out the hazards. The moment we open the chamber, we will know whether the cat is dead or alive, but it will be too late to do anything about it.

The probability of the cat's death as the result of radiation will range from 0 to 100% (or 0.0 to 1.0). This number is a meaningless statistic. However, if we use the risk matrix to assess risk, we will note that both the level of goal attainment certainty and the level of goal attainment (the goal being the puss's survival, presumably) are in fields 2, 3 and 4, and so corrective measures are required if we want to boost the cat's chances (**Figure 8**).

2.5 The theory of inertia in risk management

If we interpret risk as the influence of uncertainty on goals, we can use the theory of inertia [10] to manage risk effectively. This model is informed by the following premises and correlations:

Premise 1. The probability of positive and negative outcomes of our actions is always 50%. We have no influence over the outcome of our actions.

Premise 2. Since we cannot influence the actual event which will pass as the result of our actions, any focus on this event will be futile. The outcome for our enterprise will be the result of our preparation for the event and not of actions taken to achieve the desired outcome.

Premise 3. Preparation for all possible outcomes (negative as well as desired) should be the goal of our actions. Lack of preparation is a decision which will result in negative outcomes.

Further, we note the following correlations:

Correlation 1. Negative outcomes are the result of three classes of factors: human error, machine malfunction and other technical shortfalls and factors out of our control. The risk of negative outcomes can be minimised through multilevel monitoring and controls within the human and machine classes of factors, which would verify that decisions are taken based on sound assumptions, that actions are followed through and that machinery is kept in a working condition with timely checks, repairs and part replacements.

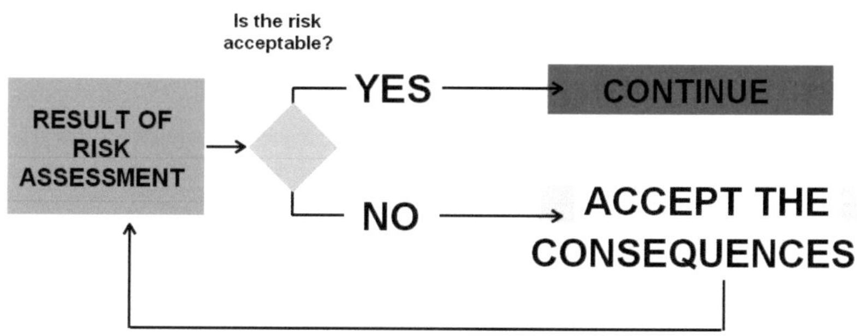

Figure 8.
Risk analysis.

The risk of negative outcomes can be minimised through multilevel monitoring and controls which would verify that decisions are taken based on sound assumptions, that actions followed through and that machinery is kept in working condition with timely checks, repairs and part replacements.

Correlation 2. If the outcome of our actions does not result from human or machine factor, we have no influence over it. In such cases we must develop contingency procedures for all outcomes beyond our influence.

3. Conclusion

The concepts of risk, danger and hazard lend themselves to statistical treatment, but the customary use of statistics for risk assessment leads to flawed conclusions, due to limited predictability of events and scenarios.

Risk management, understood as an attempt to influence the certainty of goal attainment, cannot be treated as an exercise in statistics. Instead, we ought to describe the level of certainty of goal attainment by considering costs and benefits of alternatives.

Risk assessment ought to identify the threshold parameters for accepting the costs and the level of goal attainment. The result of the assessment should identify if the goals can be achieved, to what degree and at what costs associated with adverse developments.

Every risk analysis must consider information available from documented past experiences and from identifiable future events. Their identification must be thorough and use all available tools to generate all possible scenarios.

Risk cannot be assessed without identifying the goals of the analysed process or situation. If the goals cannot be identified, we ought to limit ourselves to describing the hazards and, in the case of taking actions, the associated dangers. Additionally, we can supply historic rates of recurrence. Ultimately, statistical analysis ought not to be the primary goal of risk assessment.

When goal identification is difficult or impossible, we ought to identify the worst-case scenario and prepare for it. In the event, activities which induce danger in a hazardous situation should only be undertaken after comprehensive preparation for the identified possible effects. We ought to strive to eliminate danger arising from lack of preparation.

There are three reasons, or classes of events, which require preparation: effects of human error, effects of machine error and other events over which we have no influence. Preparation ought to comprise scenarios for each of these classes of events. The aim of the above considerations has been to suggest effective and falsifiable methods for risk assessment and for identification of hazards and dangers.

DOI: http://dx.doi.org/10.5772/intechopen.80822

Author details

Marek Różycki
M/D/R/K Trusted Adviser Group Sp. z o.o., Mikołów, Poland

*Address all correspondence to: m.rozycki@mdrk.eu

References

[1] https://www.mirror.co.uk/news/world-news/what-happened-nice-what-know-8425238 [Accessed: 10-07-2018]

[2] https://www.telegraph.co.uk/news/2016/07/15/analysis-nice-attack-shows-frances-acute-vulnerability-to-terror/[Accessed: 10-07-2018]

[3] Spignesi SJ. Catastrophe!: The 100 Greatest Disasters of All Time2004. p. 168

[4] Bhopal Information Center. Chronology. 2006, http://www.bhopal.com/chronology [Accessed: 10-07-2018]

[5] Report. http://www.unscear.org/docs/reports/2008/11-80076_Report_2008_Annex_D.pdf [Accessed: 10-07-2018]

[6] https://www.iaea.org/newscenter/news/fukushima-nuclear-accident-update-log-47 [Accessed: 10-07-2018]

[7] Williams CA Jr, Smith ML, Young PC. Risk Management and Insurance. New York: McGraw-Hill; 1964. Polish translation: "Zarządzanie ryzykiem a ubezpieczenia", Warszawa: PWN, 2002. pp. 28-29

[8] According to the ISO 31000 Definition

[9] Trimmer JD. The present situation in quantum mechanics: A translation of Schrödinger's "cat paradox" paper. Proceedings of the American Philosophical Society. 1980

[10] Różycki M. 2017 "Inertia in Procurement Risk Management" In "Sustainability and Scalability of Business: Theory and Practice" Jabłoński A. Chapter 13. Nova Publishing, New York

Chapter 11

Process of Risk Management

K. Srinivas

Abstract

Cost saving and timely performance are of utmost importance to all stakeholders who are involved in a construction project that is owner, contractor, consultant and subcontractor. The prime causes of risks in construction projects involve delay and failure to complete the work at specified cost and within the agreed time frame. Unexpected delays in construction projects are caused by internal and external environments embedding several risk factors which may occur concurrently. The cost overrun and schedule overrun not only influence the construction industry's completion of a project but can also have profound effect on the economy of a country. Even though the failure of the construction projects to get completed within the budgeted cost and time has received attention by researchers, lot more need to be researched as to what can be done to have tight leash on construction projects so that they can be brought on track. In order to meet the stiff deadlines involving complexity of construction projects, the scheduling of projects should be flexible enough to accommodate changes without negatively affecting the overall project cost and duration. This chapter deals with **Perspectives on Risk Assessment and Management Paradigms** as applicable to any project in general and construction industry in particular.

Keywords: risk management, risk assessment, risk mitigation

1. Definition of risk

Risk is defined in terms of uncertain events which may have positive or negative effect on the project objectives. Risks include circumstances or situations, the existence or occurrence of which, in all reasonable foresight, results in an adverse impact on any aspect of the implementation of the project. Various definitions of risks are presented in **Table 1**.

Sl. no	Source	Definition
1	Project Management Institute [10]	An uncertain event or a condition that if it occurs has a positive or negative effect on project objectives
2	Institute of Risk Management [7]	The combination of a probability of an event and its consequences

Sl. no	Source	Definition
3	Association of Project Management Body of Knowledge [1]	Project Risk is an uncertain event or condition, that, if it occurs, has a positive/negative effect on project objectives. A risk has a cause and if it occurs, a consequence.
4	British Standard BS IEC 62198:2001	Combination of probability of an event occurring and its consequences on project objectives
5	www.business.dictionary.com/definition/risk.html	A probability or threat of damage, injury, liability, loss, or any other negative occurrence that is caused by external or internal vulnerabilities and that may be avoided through preemptive action.
6	Fundamentals of Risk Management [9]	"A chance or possibility of danger, loss, injury or other adverse consequences" and the definition of risk is "exposed to danger." However, taking risk can also result in positive outcome. A third possibility is risk related to uncertainty of outcome.
7	Adams [8]	Risk is the probability "that a particular adverse event occurs during a stated period of time, or results from a particular challenge."
8	Philosophy of Risk [3]	Risk has been interpreted as Risk = hazard × exposure where Hazard is defined as the way in which a thing or situation can cause harm and exposure as the extent to which the likely recipient of the harm can be influenced by the hazard

Table 1.
Definitions of risk.

2. Classification of risks

Classification and definition of risks is furnished in **Table 2**.

Risk	Definitions
Pure risk	A risk which has chance of loss or no loss. ***Example***. A building may get affected by fire or not. These are best covered by insurance
Speculative risk	Involves chance of gain/loss. ***Example***. A builder may take a risk by promoting a new venture depending upon the prevailing conditions in the vicinity of proposed project, but it may bring him gain/loss.
Fundamental risk	These are external to a project and which, if they materialise, would be on a large scale and cannot be prevented. These risks are associated with major natural, economic, political or social changes and generate large scale losses. Examples are: Floods, earthquakes, fluctuation of exchange rates, etc. This risk may or may not be insurable.
Particular risk	These are project specific risks and are identified within the parameters of a project and can be controlled during the implementation of a project, e.g. quality risks, safety risks, legal risks, etc.
Source: Project Risk Management, D Van Well-Stam et al., Kogan Page Publications, 2003.	

Table 2.
Classification of risks and their definitions.

3. Introduction to risk management

Risk management is a planned and a structured process aimed at helping the project team make the right decision at the right time to identify, classify, quantify the risks and then to manage and control them. The aim is to ensure the best value for the project in terms of cost, time and quality by balancing the input to manage the risks with the benefits from such act. It is just a cost benefit analysis.

Risk management is a continuous process which is to be implemented in any project from inception to completion. However, in order to realise its full potential, risk management should be implemented at the earliest stage of a project, i.e. feasibility design and construction. Risk is an uncertain event or condition that, if occurs, has a positive or negative effect on a project's objectives. Components of risk are the probability of the occurrence of an event and the impact of the occurrence of that event. There are many sources of uncertainty in construction projects, which include the performance of construction parties, resources availability, contractual relations, etc. because of which, construction projects face problems that cause delay in the project completion time. Success of a project is measured by its ability to get completed within the budgeted cost and time. These goals are interrelated where each parameter has an impact when other parameters get affected. An accurate cost estimating and scheduling should be performed in order to meet the overall budget and time deadline of a project. As such, risk management becomes an integral part of construction management which intends to identify and manage potential and unforeseen risks during the period of implementation of the project; hence, the necessity of risk management [5].

4. Definitions of risk management

Definitions of risk management are presented in **Table 3**.

Sl. no	Definition of risk management
1	**https://en.wikipedia.org/wiki/Risk_management** Risk Management is the identification, evaluation, and prioritization of risks followed by coordinated and an economical application of resources to minimise, monitor, and control the probability or impact of unfortunate events [6] or to maximize the realization of opportunities.
2	**Nadeem Ehsan et al., 2012** Risk Management in a project involves the identification of influencing factors which could have negative impact on the the cost, schedule and quality objectives of the project and quantification of impact of potential risk and implementation of mitigation measures to minimise the potential impact of risk
3	**Bahamid et al., 2017** Risk Management is defined as organized and comprehensive method tailored towards " organizing", "identifying" and "responding" to risk factors in order to achieve project goals.
4	**www.stakeholdermap.com/risk/risk-management-construction** Risk Management in construction consists of planning, monitoring and implemeting the measures needed to prevent exposure to risk. To do this, it is necessary to identify the hazards, assess the extent of risks, provision of measures to control the risks and to manage residual risk
5	**www.vp-projects.kau.edu.sa** Risk management is a systematic method of identifying, analysing, treating and monitoring the risks that are all involved in any activity/ process and is a systematic method that minimises the risks which may be an impediment to attainment of objectives

Sl. no	Definition of risk management
6	**Cleden** [4] Risk is exposure to the consequences of uncertainty. In a project context, it is the chance of something happening that will have an impact upon objectives. It includes the possibility of loss or gain, or variation from a desired or planned outcome, as a consequence of the uncertainty associated with following a particular course of action. Risk thus has two elements: the likelihood or probability of something happening, and the consequences or impacts if it does.
7	**Project Risk Management, D Vanwell-Stam, Kogan Page India publications, 2004** The entire set of activities and measures that are aimed at dealing with risks in order to maintain control over a project
8	**www.gpmfirst.com/risk.** ***management-construction*** Risk Management is a means of dealing with uncertainty – identifying sources of uncertainty and the risks associated with them, and then managing those risks such that negative outcomes are minimized (or avoided altogether), and any positive outcomes are capitalised upon.
9	**Risk Management in Construction Projects by NICMAR** [2] Risk Management is the planned and structured process of bringing the project team make the right decisions at the right time by identifying, classifying and quantifying the risks and then for managing and controlling them,
10	**Dr Patrick et al., 2006 retrieved from feaweb.aub.edu.lb** Risk Management is "a systematic way of looking at areas of risk and consciously determining how each should be treated. It is a management tool that aims at identifying sources of risk and uncertainty, determining their impact, and developing appropriate management responses"
11	**http://economictimes.indiatimes.com/definition/risk-management** Risk Management refers to the practice of identifying potential risks in advance, analysing them and taking precautionary steps to reduce/curb the risk.

Table 3.
Definitions of risk management.

5. Importance of risk management in construction projects

Construction projects are extremely complex and fraught with uncertainty. Risk and uncertainty can potentially have damaging consequences for the construction projects. Hence, risk analysis and risk management has come to be a major feature of the project management in construction projects. Construction projects are unique, inherently complex, dynamic and risks emanate from multiple sources. The interests of individuals and organisations who are actively involved in a construction project may be positively or negatively affected depending upon the course which a project takes from concept to completion. Multiple stakeholders with varied experience and skills have different expectations and interests in the project which creates problems for smooth execution of the project. Risk management is a concept which many construction companies have never thought of, despite the fact that, the risks can be better controlled if they are identified in the first instance and a well-structured mitigation mechanism is in place. Risk management helps the key project participants namely the client, contractor/developer, consultant and supplier to meet their commitments and to minimise negative impacts on construction project performance in relation to cost, time and quality objectives. Success of a construction project is associated with three aspects of time, cost and quality outcomes.

Successful commissioning of any project, necessarily calls for sound planning on various fronts and getting the project executed in a competent manner. An organisation executing a project would have to reckon with the various risks to which the project may be exposed to and these have to be managed effectively. The construction industry, being vulnerable is potentially more prone to risks and uncertainties than any other industry. The process of taking a project from the conceptual stage to its final completion and putting into operation is quite complex and entails painstaking process at every stage. Construction industry is highly fragmented in that each of its participants—designers, constructors, planners, suppliers, etc. can be highly skilled in their own area and yet there is no clear perspective as to how all the players can come on the same platform for achieving the objectives.

Construction industry is also dependent on quality of its people rather than technology. The increasing technological complexity and more complex interdependencies and perpetual shortage of resources namely materials, equipment, technical/supervisory staff, finance, etc. calls for a comprehensive risk management framework which will insulate the risks of the participants to a great extent.

Given the nature of the construction sector, risk management is an extremely important process. It is most widely used in such of those projects where susceptibility to risks is very high and is characterised by planning, monitoring and controlling the risks in a more structured and formal manner. The most efficient method of identifying the risks is to study a project of similar size which was executed in the recent past which gives an insight into the failure/success of the project. In order to be sure that the project objectives are met, the portfolio of risks associated with all stakeholders should be considered across the project life cycle (PLC). In later stages, risk management when applied systemically helps to control those critical elements which can negatively impact project performance. Keeping track of identified threats will result in early warnings to the project manager if any of the objectives, time, cost or quality, are not being met. There are a plethora of risks which are to be identified in the construction industry and which can be faced in each construction project at any point of time regardless of its size and scope. Frequent change in scope is one of the major risks in any construction project. If revised scope or design is implemented, it can have effect in the form of additional resources of time and cost. Early project completion may be as troublesome as delays in a schedule. Completing too early which may be a result of insufficient planning or design problems can lead to a low quality of final product and increased overall cost. Thus it is important to keep a balance in the concept of time–cost-quality trade-off, which more widely is becoming an important issue for the construction sector. Risks may vary depending on the project scope, types and are to be treated accordingly.

6. Risk management process

Risk management process is shown in **Figure 1**.

6.1 Identification of risks

Risk identification, the first step in the risk management process is usually informal and is performed in various ways, depending on the organisation and the project team. Identification of risks relies mostly on past experience and study of similar executed projects. This being a preliminary stage, a combination of tools and techniques may be used to identify the risks in any project. Here are many methods that fit specific types of challenges and projects especially at identification stage. Risks and threats may be difficult to eliminate, but when they have been

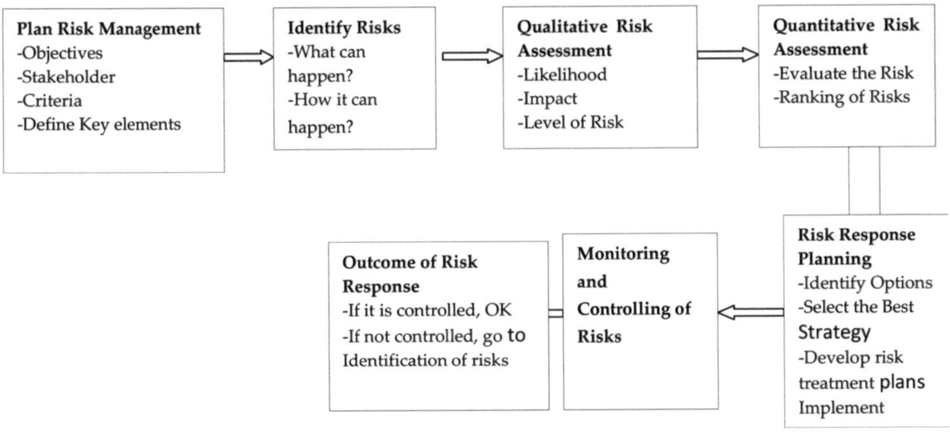

Figure 1.
Risk management process.

identified, it becomes easy to take actions and have control over them. Risk management will be more effective if the source of the risks have been identified and allocated before any problems occur. The main purpose of risk management is that the stakeholders should prepare for potential problems that can occur unexpectedly during the course of a project. Risk management will not only facilitate anticipating problems in advance, but also preparing oneself for the potential problems that may occur unexpectedly. Handling potential threats is not only a way to minimise the losses within a project, but also a way to transform risks into opportunities which can lead to economic and financial profitability. The purpose of identifying risks is to obtain a list of risks which has got the potential to have a cascading effect on the progress of project and different techniques are applied for managing/mitigating the same. In order to find all potential risks which might impact a specific project, different techniques are applied. The project team should use a method they are familiar with so that the exercise will be effective. Effective identification of risks is the first step to a successful risk management.

6.1.1 Risk identification techniques

Tools and techniques for risk identification are presented in **Table 4**.

Parameter	Methodology
Documentation reviews	A structured review of project documentation, study of history of execution of similar projects and quality of plans as well as the consistency between those plans and project requirements/ assumptions would be an indicator of risks in the project
Information gathering techniques	• Brainstorming • Delhi technique • Checklist analysis • Cause and effect diagram • Questionnaires • SWOT analysis • Expert judgement

Table 4.
Identification of risks: tools and techniques.

Various risks that confront a construction industry are not limited to and include financial, economical, political, legal environmental, technical, contractual, planning/scheduling, design, quality operational labour, stakeholder safety and security, logistics and construction.

6.2 Risk assessment

Risk assessment is the second stage in the risk management process where collated data is analysed for potential risks. Risk assessment is described as short listing of risks starting from low impact highest impact on the project, out of all threats mentioned in the identification phase. Risk assessment consists of qualitative risk assessment and qualitative risk assessment.

6.2.1 Qualitative risk assessment

This involves registration of identified risks in a formal manner. A risk register is used for formalising this process which is not limited to the following

- Classification and reference
- Description of the risk
- Relationship of the risk to other risks
- Potential impact
- Likelihood of occurrence
- Risk response/mitigation strategy
- Allocation of risks to stakeholders.

6.2.1.1 Classification and reference

Classification is an aid to identifying the source of risk. Examples are furnished below (**Table 5**).

Referencing refers to unique reference number given for each of the identified risks.

6.2.1.2 Description of the risk

This involves giving a brief description of the risk. The description must be unique in order to avoid confusion with similar risks in the risk management process.

Risk	Classification
Environmental	Site conditions, health and safety issues at site
Contractual	Client, contractor, sub-contractor, etc.
Design	Planning permission, preliminary and detailed design, etc.

Table 5.
Classification of risk and its reference.

6.2.1.3 Relationship to other risks

In any project, it is extremely rare that any activity is independent of activities which occurs concurrently or consequentially and this will always be the case for risks also for successful implementation of risk management,

6.2.1.4 Potential impact (I)

Impact of risk on a project is measured in terms of cost and quality. Since this assessment is done at an early stage of the project, information may not be available to accurately predict the impact of risk on the project. At this stage, the risk is classified suitably and accordingly high impact risks are to be given more fundamental consideration than that of medium/low/negligible risks by ranking the impact of risks on a scale of 1 (low) to 10 (high).

6.2.1.5 Likelihood of occurrence (P) and calculation of risk factor (RF)

Based on intuition and experience, the likelihood of occurrence (P) of risks and its impact (I) is to be given on a suitable scale ex. 1–10 (1 refers to low probability and 10 refers to high probability). The risk factor for each of the identified risks is calculated by the formula $RF = P + I - (P*I)$ (where the values of P and I are brought on a scale of 0–1 by dividing the values with 10).

6.2.1.6 Risk response/mitigation strategy

This action is taken to reduce, eradicate or to avoid the identified risks. The most common among the risk mitigation methods are risk avoidance, risk transfer, risk reduction and risk sharing. Based on the competency in handling the risks, the identified risks are allocated to respective stakeholders who will be responsible for addressing those risks.

6.3 Quantitative risk assessment

This risk assessment is normally taken for such of those risks which are classified are high/critical/unmanageable as per the qualitative risk assessment. The purpose of this assessment is to find the amount of contingency to be inserted in the estimate for the risks undergoing this assessment so that in case the risks occur, there would be sufficient budgeted amount to overcome the extra expenditure.

Quantitative methods need a lot of analysis to be performed. This analysis should be weighed against the effort and outcomes from the chosen method. Complex and larger projects require more in depth analysis as compared to projects which are small in size. The purpose of carrying out quantitative analysis is to estimate the impact of a risk in a project in terms of scope, time, cost and quality. The suitability of this analysis is more for medium and large projects as these projects have more complex risks as compared to smaller projects.

The detailed quantitative assessment of risk is the one which is identified as risk analysis. In undertaking quantitative assessment, the potential impact of risks in terms of time, cost and quality is quantified. While preparing the estimate, it is generally split into two distinct elements, namely (1) base estimate of those items which are known and a degree of certainty exists and (2) contingency allowance for all uncertain elements of a project. Historically, contingencies have been calculated on a rule of thumb basis varying from 5 to 10% on risk-free base estimate. By adopting risk management approach, contingencies are set up to reflect realistically the risks that are inherent in the project. When used correctly, contingency allowances ensure that expenditure against risks is controlled. The methods for quantitative risk assessment are described below.

6.3.1 Scenario technique: Monte Carlo simulation

The Monte Carlo method is based on statistics which are used in a simulation to assess the risks. This is a statistical technique whereby randomly generated data is used within predetermined parameters and produce realistic project outcomes. The overall project outcome is predicted by randomly simulating a combination of values for each risk and repeating the calculation a number of times and all outcomes are recorded. After completing the simulations required, the average is drawn from all of the outcomes, which will constitute the forecast for the risk. It is important to realise that parameters and appropriate distribution within which the random data is simulated is itself a series of subjective inputs. Accurate and realistic project outcomes will not be generated if inaccurate parameters are set. Different scenarios are generated by simulation are used for forecasting, estimations and risk analysis. Data from already executed projects is normally collected for simulation purpose. The data for variables is presented in terms of pessimistic, most likely and optimistic scenarios depending upon the risks encountered, i.e. pessimistic value means lot of risks and optimistic value means least risks. The result from this method is a probability of a risk to occur is often expressed as percentage. The most common way of performing the Monte Carlo simulation is to use the program Risk Simulator Palisade Software, where more efficient simulations can be performed.

6.3.2 Modelling technique: sensitivity analysis

This is a method used to demonstrate the variable impact on the whole caused by a change in one or more element or risk. It is used to test the robustness of choices made where rankings have been established, particularly when those rankings are considered to be marginal. It can identify the point where variation in one parameter will affect decision making. A typical method for carrying out sensitivity analysis is by use of a spider diagram which shows the areas in the project which are the most critical and sensitive The higher the level of uncertainty a specific risk has, the more sensitive it is concerning the objectives. In other words, the risk events which are the most critical to the project are the most sensitive and appropriate action needs to be taken (Heldman, 2005). Disadvantage with this analysis is that the variables are considered separately, which means that there is no connection between them (Perry, 1986 and Smith et al.. 2006). The method requires a project model in order to be analysed with computer software. According to Smith et al. (2006), the project stands to be benefited if the analysis is carried out in the initial phases of a project in order to focus on critical areas during the execution of the project.

6.3.3 Decision tree

Decision tree analysis is commonly used when there is sequence of interrelated possible courses of action and future outcomes in terms of time and cost. This method of analysis is commonly used when certain risks have an exceptionally high impact on the two main project objectives, i.e. time and cost. Where probabilities and values of potential outcomes are known or can be estimated, they are used for quantification to provide a more informed basis for decision making. Each decision process expected value (EV) which forms the basis for decision making process. A sample problem on decision tree is given in **Table 6**.

Method	Design time (months)	Construction period in months and probabilities	Total time (construction period + design time) (months)
Construction management	2	15 (0.6) = 9 18(0.4) = 7.2 **Total 16.2 months**	18.2
Design and construct	3	12(0.3) = 3.6 **Total = 13.8 months** 14(0.5) = 7 16(0.2) = 3.2	16.8
Traditional method	8	10(0.3) = 3 12(0.7) = 8.4 **Total 11.4 months**	19.4

Table 6.
Problem on decision tree.

This can be depicted in the form of decision trees and the expected value (EV) in terms of time for each of the three scenarios is furnished. The least of this i.e. construction management will be preferred since it consumes less time.

6.3.4 Multiple estimating using risk analysis

Multiple estimating using risk analysis (MERA) attempts to provide a range of estimates. These are presented as risk free base estimate, average risk estimate (ARE) and maximum likely risk estimate (MLRE). ARE is the sum of risk free base estimate and average risk allowance and MLRE is the sum of ARE and maximum risk allowance.

MERA attempts to finds a level i.e. the estimate that has a 50% chance of being successful. This is known as average risk estimate (ARE) which is found out by multiplying the average allowance with average probability of occurrence. Maximum risk allowance is found out by multiplying the maximum allowance with maximum probability of occurrence of that risk. This is added to ARE to get MLRE which is the estimate that has 90% chance of not being exceeded.

6.3.5 Quantitative risk assessment: outputs

The output of quantitative risk assessment is presented in **Table 7**.

Parameter	Outputs
Probabilistic Analysis of project	Estimates are made of potential project schedule and cost outcomes listing the possible completion dates and costs with their confidence levels. This output is described as cumulative distribution and also risk tolerances for permitting quantification of cost and time contingency reserves. Contingency reserves bring the risk of overshooting stated project objectives to acceptable levels to the organisation
Prioritised list of quantified risks	This list includes risks that pose the greatest threat or present the greatest opportunity in a project. These risks also have the greatest impact on cost contingency
Trends in quantitative risk analysis results	As the risk analysis is repeated, a trend becomes apparent that leads to conclusions affecting risk responses, Historical information on project's schedule, cost, quality and performance reflects new insights gained through quantitative process. This takes the form of quantitative risk analysis report.

Table 7.
Quantitative risk assessment: outputs.

6.4 Risk response planning

The risk response will be in the form of mitigation by adopting necessary strategies in respect of positive and negative risks which is furnished below (**Tables 8–10**).

6.5 Monitoring and controlling risks: inputs

Inputs to monitoring and controlling of risks are presented in **Table 11**.

6.5.1 Monitoring and controlling risks: tools and techniques

Tools and Techniques for monitoring and controlling risks are furnished in **Table 12**.

Risk mitigation strategy	**Description**
Risk avoidance	Risk avoidance involves changing the project management plan to eliminate the threat entirely. The project manager may isolate the project objectives that are in jeopardy. **Examples:** (a) Extending the schedule of an activity; (b) Changing the strategy or reducing the scope of work; (c) Changes in clauses of contract regarding abnormal price rise of any material or dealing with extra quantum of work.
Risk transfer	Risk transfer requires shifting some or all of the negative impact of a threat along with ownership of the response to a third party. Examples are • **Risk transferred to Consultant** Design Risk, technical Risk and Foundations for all major structures • **Risk transferred to Insurance Company** Security of materials at site, Fire Hazards, Boiler operations, safety of electrical rooms, loss in Turbines and Generators, Unforeseen Risks, etc.
Risk reduction	Risk reduction implies reduction in the probability and consequence of an adverse risk event to be within acceptable threshold limits. Conducting detailed tests or choosing a more stable supplier are some examples. Risk reduction is adopted where the resultant increase in costs is less than the potential loss that could be caused by the risk being mitigated. **Examples are**: • Preparedness to tackle any natural disaster • Detailed site investigation where adverse ground conditions are known to exist but the full extent is not known. A detailed ground investigation was performed upon which an estimate was prepared. • Contingency planning • Removal of engineering/structural barriers • Strengthening the quality assurance procedures • Paying higher amount than recommended by Govt for land acquisition • Design as per standards
Risk acceptance	This strategy is adopted when it is not possible to eliminate all risks from a project. This strategy indicated that the project team had decided not to change the project management plan or is unable to identify any other suitable response strategy. This requires no action except to document the strategy leaving the project team to deal with risks as they occur

Table 8.
Strategies for mitigating negative risks.

Risk mitigation strategy	Description
Exploit	This strategy is selected for risks with positive impacts where the organisation wishes to ensure that the opportunity is realised. This strategy seeks to eliminate the uncertainty associated with a particular risk by ensuring that the opportunity is exploited. **Examples** are assigning the most talented resources of the organisation to the project to reduce the time for completion or providing at a lower cost than originally planned
Share	Sharing a positive risk involves allocating some or all of the ownership of the opportunity to a third party capable of capturing the opportunity for the benefit of the project. Risk sharing, joint ventures, etc. are examples of this strategy
Enhance	This strategy is used to enhance the positive impact of an opportunity. Identifying and maximising key drivers of risks may increase their probability of occurrence. **Examples** are adding more resources to an activity for completing it before scheduled time
Accept	Accepting an opportunity means willing to take advantage if it comes along, but not pursuing it actively.
Contingent response strategies	Some responses are designed for implementation only if certain events occur. It is appropriate for the project team to prepare a contingency response plan that will be executed under certain predefined conditions if there will be sufficient warning to implement the plan
Expert judgement	Expert judgement is from knowledgeable individuals pertaining to the actions to be taken on a specific and a defined risk.

Source: Project Risk Management, Van Well Stam et al. 2008.

Table 9.
Risk mitigation strategies for positive risks/opportunities.

Sl. No	Contents
1	Identified risk. Their descriptions, areas of project affected, their causes and how they affect project objectives
2	Risk owners and assigned responsibilities
3	Prioritised list of project risks based on the outputs from quantitative analysis reports
4	Agreed upon response strategies and specific actions taken to implement the strategy
5	Triggers, symptoms and warning signs of risks occurrence
6	Fallback plans as a reaction to a risk that has occurred and primary response proved to be inadequate
7	Contingency reserves to be calculated based on quantitative risk analysis of the project and the threshold risk of the organisation.

Table 10.
Contents of risk response: outputs.

Parameter	Inputs
Risk register	The key inputs to risk register includes identified risks and owners of risk, agreed upon risk responses, specific actions to be implemented, symptoms/warning signs of any risk, residual/secondary risks, list of low priority risks and contingency measures in terms of time/cost
Risk management plan	The risk management plan should contain risk tolerances, assignment of manpower including bearer of risk, time and other resources to project risk management
Work performance information	Work performance information related to various performance results is to be quantified in terms of deliverable status, schedule progress and costs incurred.
Performance reports	Performance reports will be analysed for variance analysis, earned value data and forecasting the likely date of completion of project

Source: [10].

Table 11.
Monitoring and controlling risks—inputs.

Parameter	Explanation
Risk reassessment	Monitoring and controlling of risks will result in identification of new risks, reassessment of current risks and closing of risks that are not a threat to project. Project risk assessment is to be performed regularly. The frequency and depth of assessment depends on how the project progresses relative to the objectives.
Risk audits	The purpose of risk audits is to examine and document the effectiveness of risk responses in dealing with identified risks and their root causes as well as effectiveness of the risk management process meetings. A separate risk audit meeting may be held or it may be included in routine project review meetings.
Earned value analysis	Variance analysis is done by comparing the planned results with actual. Trends in execution of a project are reviewed using performance information and based on earned value analysis, deviation from cost and schedule targets is determined which may indicate potential impact of threat/opportunities.
Technical performance measurement	This measures technical accomplishments during project execution and will help in forecasting degree of success in achieving the project scope and it may expose the degree of technical risk faced by the project
Reserve analysis	This compares the amount of contingency reserves available at any time to the amount of risk remaining in the project to determine whether the reserves are adequate
Status meetings	Project risk management should be an agenda item at all status review meetings. The amount of time for any item will depend on risks that have been identified, their priority and difficulty of response. Frequent discussions about risk make it more likely that concerned stakeholders will identify risks and opportunities.

Source: [11].

Table 12.
Monitoring and controlling risks: tools and techniques.

7. Benefits with risk management

- To maximise the efficiency of risk management, the risk management process should be continuously developed during the entire project.

- The benefits from risk management finally go to the stakeholders involved. A clear understanding and awareness of potential risks in the project contributes to better management of risks by suitable mitigation techniques. Another benefit of working with risk management is increased level of control over the whole project and more efficient problem solving processes which can be supported on a more genuine basis

- Risk management when conducted effectively, reduce sudden surprises. The advantage with risk management is that the stakeholders are aware as to the risk that they have to bear among all the risks that have been identified in a project and can prepare themselves accordingly, should any eventuality occur. No doubt, this formal exercise may translate into extra cost for an activity, but if taken in holistic manner, the benefits will far outweigh the costs. This has another advantage in that there is no passing of buck as risks are either shared/retained or transferred depending upon the ability of the stakeholder to handle the risk. The three approaches to risk management are normally risk natural firm which does not invest much in risk management but is still aware of important risk, risk averse firm where no investments are made and the last one is risk seeker wherein the organisation is prepared to face all risks and is often called gambler. The outcome of the objectives of project naturally depends upon the path adopted by the firms in their approach to risk management.

8. Conclusions

- The fact that there are manifold risks which can be identified in any construction project is explained by their size and complexity. Bigger the project is, the larger the number of potential risks that may be faced.
- Occurrence of risk is stimulated by several factors. Most often the risks faced in any project are financial, environmental (surrounding location of project and overall regulations), time, design and quality. The technology used for construction and the internal environment also contributes to risk which can have substantial bearing on the outcome of a project.
- Risks are directly proportional to complexity of a project. Bigger and more complex a project is, the more resources are required to complete it. In spite of identifying all potential risks, there might be more potential threats. Therefore, the project team should not solely focus on management of those identified risks but also be alert for any new potential risks which may arise during execution.
- Risk management is a tool for managing risks in a project and a project manager should be prepared for managing uncertainties not included in a risk management plan.
- Effective management of risky project demands rapid and realistic predictions of alternative courses of action and positive decision making and requires flexible attitudes and procedures.
- Perception of severity and frequency of occurrence of risk is to be done in tandem between the stakeholders. This will eliminate lot of unnecessary correspondence as well as misunderstanding and friction between the stakeholders
- Insurance is just one aspect of risk mitigation and it cannot absorb all the risks. Insurance is project specific and it should be taken as per the needs of client/contractor. Other ways of risk mitigation needs to be explored.
- Adoption of good project management practices like proper planning and implementation, willingness of stake holders to share the risks in the project is essential for success of a project
- Executing a complex project requires meticulous planning, i.e. planning to the smallest details, and this can be achieved through concerted dedication from the concerned stakeholders.
- Risks are to be thoroughly studied and understood before bidding for the project.
- Special care should be taken regarding the seasonal variation of labourers, so that the construction activities does not get delayed due to shortfall in manpower resources during execution, which can have adverse effects on cost and time
- Proper risk allocation techniques should be framed between the stakeholders so that in the event of occurrence of a risk, this will eliminate doubts as to which stakeholder should address the risk

- Given its complexity, risk management is a very important process in construction projects. It is most widely used in those projects which exhibits high level of uncertainty. Formal planning, assessment and monitoring/control process characterises risk management in such projects.

- Risk management procedures should be initiated in the early stages of the project where planning and contracting of work, together with the preliminary capital budget are being chalked out. In later stages, Risk management applied systemically, helps to control those critical elements which can have negative impact on project performance.

- Keeping track of identified threats, will result in early warnings to the project manager if any of the objectives, time, cost or quality, is being met or not.

- Risks in complex construction projects can be mitigated by entering into various agreements like execution, operation/maintenance, etc.

- Proper risk strategy formulation and research is necessary based on real life experiences so that identification of potential risks and providing solutions can produce effective and efficient risk strategies to overcome impacts of risk events.

- Risk identification is the first step in the risk management process. It means that the identification of risks which is informal relies mostly on past experience of similar executed projects and that of advice from experts. There are a good number of methods for identifying the risks in a project and a combination of methods may be used for identification of risks in a project.

- Handling potential threats is not only a way to minimise losses within the project, but also a way to transform risks into opportunities, which can lead to economical profitability and finally, .it is suggested that if risks are given due care at all stages of the project, stakeholders will be showered with manifold benefits subsequent to commissioning of project

Author details

K. Srinivas
National Institute of Construction Management and Research, Pune, Maharashtra, India

*Address all correspondence to: ksrinivasap@gmail.com

References

[1] Association of Project Management Body of Knowledge. Retrieved from: http://www.cs.bilkent.edu.tr/~cagatay/cs413/PMBOK.pdf

[2] Banaitiene N, Banaitis A. Risk Management in Construction Projects. Retrieved from: https://cdn.intechopen.com/pdfs/38973/IntechRisk_management_in_construction_projects_pdf

[3] Chicken JC, Posner T. The Philosophy of Risk, Vol. 21. London: Thomas Telford; 1998

[4] Cleden D. Managing Project Uncertainty, 1st ed. Great Britain: Gower Publishing Ltd; 2009

[5] Cooper D, Grey S, Raymond G, Walker P. Project Risk Management Guidelines—Managing Risk in Large Projects and Complex Procurements. Wiley Publishers; 2005. ISBN 9780470022825

[6] De Marco A, Thaheem MJ. Risk analysis is construction projects—A practical selection methodology. American Journal of Applied Sciences. 2014;**11**(1):74-84

[7] Institute of Risk Management. Retrieved from: https://www.theirm.org/media/886059/ARMS_2002_IRM.pdf

[8] Adams J. 1995. Retrieved from: http://www.john-adams.co.uk/wp-content/uploads/2017/01/RISK-BOOK.pdf

[9] Hopkin P. Fundamentals of Risk Management. 5th ed. Great Britain: Kogan Page Limited; 2013

[10] Project Management Institute. Guide to Project Management Body of Knowledge. 4th ed. USA: PMI; 2008. ISBN 978-1-933890-51-7

[11] Risk Management. India: School of Distance Education, National Institute of Construction Management and Research